MALABAR
FLORIDA 32950

GRAPHICAL EVOLUTION

WILEY-INTERSCIENCE
SERIES IN DISCRETE MATHEMATICS

ADVISORY EDITORS

Ronald L. Graham
AT&T Bell Laboratories, Murray Hill, New Jersey

Jan Karel Lenstra
Mathematisch Centrum, Amsterdam, The Netherlands

Graham, Rothschild, and Spencer
RAMSEY THEORY

Tucker
APPLIED COMBINATORICS

Pless
INTRODUCTION TO THE THEORY OF ERROR-CORRECTING CODES

Nemirovsky and Yudin
PROBLEM COMPLEXITY AND METHOD EFFICIENCY IN OPTIMIZATION
(Translated by E.R. Dawson)

Goulden and Jackson
COMBINATORIAL ENUMERATION

Gondran and Minoux
GRAPHS AND ALGORITHMS
(Translated by S. Vajda)

Fishburn
INTERVAL ORDERS AND INTERVAL GRAPHS: A STUDY OF PARTIALLY
ORDERED SETS

Tomescu
PROBLEMS IN CONBINATORICS AND GRAPH THEORY
(Translated by Robert A. Melter)

Palmer
GRAPHICAL EVOLUTION: AN INTRODUCTION TO THE THEORY OF
RANDOM GRAPHS

GRAPHICAL EVOLUTION

an introduction to the theory of random graphs, wherein the most relevant

PROBABILITY MODELS

for graphs are described together with certain

THRESHOLD FUNCTIONS

which facilitate the careful study of the structure of a graph as it grows and specifically reveal the mysterious circumstances surrounding the abrupt appearance of the

UNIQUE GIANT COMPONENT

which systematically absorbs its neighbors, devouring the larger first and ruthlessly continuing until the last

ISOLATED VERTICES

have been swallowed up, whereupon the Giant is suddenly brought under control by a

SPANNING CYCLE

The text is laced with challenging

EXERCISES

especially designed to instruct, and is accompanied by an

APPENDIX

stuffed with useful formulas that everyone should know.

EDGAR M. PALMER

Department of Mathematics
The Michigan State University

A WILEY-INTERSCIENCE PUBLICATION

JOHN WILEY & SONS

New York · Chichester · Brisbane · Toronto · Singapore

A verbal contract isn't worth the paper it's written on.

SAMUEL GOLDWYN

Library of Congress Cataloging in Publication Data:

Palmer, Edgar M.
   Graphical evolution.

   (Wiley-Interscience series in discrete mathematics)
   Subtitle: An introduction to the theory of random
graphs, wherein the most relevant probability models for
graphs are described together with certain threshold
functions which facilitate the careful study of the
structure of a graph as it grows and specifically reveal
the mysterious circumstances surrounding the abrupt
appearance of the unique giant component which systematically absorbs its neighbors, devouring the larger first
and ruthlessly continuing until the last isolated vertices
have been swallowed up, whereupon the giant is suddenly
brought under control by a spanning cycle. The text is
laced with challenging exercises especially designed to
instruct, and is accompanied by an appendix stuffed with
useful formulas that everyone should know.

   "A Wiley-Interscience publication."
   Bibliography: p. 163
   Includes indexes.
   1. Random graphs.   I. Title.   II. Series.
   QA166.17.P35   1985   511'.5   84-25695
   ISBN 0-471-81577-2

Printed in the United States of America

10 9 8 7 6 5 4 3 2

It takes two hands to handle a whopper.
<small>FISHERMAN'S PROVERB</small>

To

Paul Erdös and Alfred Rényi,

*who founded the theory of evolution of random
graphs and discovered
many beautiful theorems therein.*

PREFACE

*If the people don't want to come out to the park,
nobody's going to stop them.*

YOGI BERRA

The theory of random graphs may have its roots in the classical enumeration
theory initiated by Cayley as well as the vague regions of probabilistic folklore,
but the fundamental papers, [ErR59] and [ErR60], of Paul Erdös and Alfred
Rényi still seem to have come forth spontaneously and without any warning.
They covered so much ground so thoroughly and found so many beautiful
results that the expected readership for such subject matter was caught by
surprise. Evidently a period of about ten years was required for adequate
digestion of these discoveries. A glance at Karoński's bibliography [Ka82]
shows that only a handful of papers were written about the subject in the
1960s, whereas several hundred articles have appeared since 1970! We shall see
that several important random graph problems have been solved in just the last
few years, and there are still plenty of interesting questions left unsolved. So
one would expect interest and research activity in this area to continue,
especially when it is related to computer science or theoretical chemistry where
the most interesting applications may be.

It seemed to me that my graduate students in both mathematics and
computer science ought to have an opportunity to find out about a topic that
had a 25 year period of development, strong indications of research momen-
tum and excellent potential for applications. Therefore this book first took the
form of a set of notes for my seminar participants who wanted to learn about
random graphs. The notes were subsequently revised for use in part of my
graduate course in graph theory and combinatorics. Naturally the ability of the

students in our seminars and classes here at Michigan State University has had some influence on the level of difficulty of the text. Year after year we have seen a steady procession of outstanding students, including hoards of Merit Scholars, Putnam Scholars, Churchill Scholars, Rhodes Scholars and winners of the prestigious Alumni Distinguished Scholar awards. This year, for example, we have for the third consecutive year a winner of a Rhodes Scholarship for women. A fair number of these people are enrolled in mathematics or computer science and hence, much to our delight, in combinatorics courses. Nevertheless, much effort has been made to provide an easy introduction to the subject for first-year graduates and advanced undergraduates. Many of the exercises require only a routine application of a definition or theorem. Others can be handled by copying a method carefully illustrated in the text. A few in the later chapters ask for the verification of simple steps omitted in proofs. None are terrifically long or laborious. Nor do they depend on some mystical flash of brilliance that occurs only once a month.

The text begins with a brief review of enumeration theory. Famous results of Cayley, Otter, Pólya, Redfield and Wright are mentioned without proof. Then in Chapter 2 we indicate the relationship between enumeration and the theory of random graphs with illustrations for connected graphs. Next we review basic probabilistic concepts as used for graphs while we continue to explore the property of connectivity. Chapter 3 provides a fairly thorough exposition of the method of discovery of threshold functions. Here, exercises 3.1.4, 3.1.5 and 3.2.4 are especially suitable to practice the technique.

At this point some of the main results from the pioneering papers [ErR59] and [ErR60] of Erdös and Rényi can be sketched. Most of our attention in Chapter 4 is concentrated on the amazing double-jump threshold, where the random graph undergoes rapid and drastic structural modification. But we also follow up our earlier work in connectivity with sharper results.

Several important topics have been selected for closer study in Chapter 5. Erdös and Rényi made significant contributions to the determination of the degree distribution of a random graph, and these have been refined and extended by Ivčhenko and Bollobás. The chromatic number was investigated not only by Erdös and Rényi but also by the teams of Grimmett and McDiarmid, Erdös and Spencer, Erdös and Bollobás, and Garey and Johnson. Matula was the first to discover the remarkable properties of the clique number of a random graph, and his startling investigation is traced in some detail. Then we turn to the probabilistic analysis of theorems in graph theory focusing especially on those that provide sufficient conditions for hamiltonicity. Finally the latest results of Fenner and Frieze for connectivity and hamiltonicity in Model C are brought out. All these topics are of considerable interest to computer scientists because of their role in the study of algorithms.

Chapter 6 deals with probability models whose investigation perhaps owes more to enumeration theory than to probability theory. But the subject matter is so interesting, so new and so closely related to the main theme of the book that it had to be included. First we generate random trees and graphs with the algorithms of Dixon, Nijenhuis and Wilf. Then there are important theorems on random regular graphs found by several investigators, including Bollobás, Fenner, Frieze, McKay and Wormald. Finally we come to the ultimate refinement of Cayley's enumeration of trees in which the simultaneous distribution of the vertices by degree and orbit size is determined. There were many contributors to this project, and those most responsible for applying the finishing touches were Bailey, Kennedy, Robinson and Schwenk.

Since we have cut across the boundaries of several disciplines, it has been necessary to provide some background material in the Appendix. There we have put all the relevant combinatorics, graph theory and probability theory that the reader should need. A few important definitions are included in both the Appendix and the text in order to maintain a smoother flow.

To emphasize the methods and motivate new material we sometimes have theorems following proofs. Furthermore, it seems to be the nature of this theory that the proofs are elementary but often far from simple. Therefore some proofs that follow a familiar pattern already demonstrated have been omitted, and in order to provide more complete explanations we do not always prove the strongest possible result. But a student who has studied this book should be well prepared to tackle all the good stuff in the literature that could not be included. The best place to begin is [ErR60], which is loaded with hot items that we have omitted. For more recent work there is the review article by Karoński [Ka82], which contains about 250 references, and the survey by Bollobás [Bo81a] with about 80 references. Mentioned there and also included in the references of this text are a number of articles on applications. For example, see Cohen [Co-U] (obtainable from that author) and Kennedy ([Ke81], [Ke83]) for chemistry, Karp [Ka79] for computer science, Grimmett [Gr83] for percolation theory and Spencer [Sp78] for coding theory. A wider treatment of probabilistic methods applied to combinatorics can be found in the book by Erdös and Spencer [ErS74].

There are many people to thank for their contributions to the organization and content of this work. Among them are the following graduate students who attended my seminar: Xiabo Li, Rick Hoffman, Jon Awbrey, Ab Manning, Kathy McKeon, and Garry Johns. My colleague Robert W. Robinson gave me much encouragement and many helpful suggestions.

I would also like to thank Ronald L. Graham for selecting this work to appear in the Wiley-Interscience Series in Discrete Mathematics. His enthusiastic support was greatly appreciated.

I'm very grateful to the secretarial staff of Michigan State University for their patience in typing quickly and accurately from my rough drafts. They include Cathy Friess, Tammy Hatfield, Kathy Higley, Cindy Smith and Sharon Tice.

Special thanks for comfort and joy go to my first wife, Jane, and favorite daughters, Amelia and Angela, all survivors of the "Auberges de Jeunesse."

Of course we are all indebted to Erdös and Rényi for discovering this beautiful theory and to the other experts listed in this text and the bibliographies of Bollobás and Karoński. It is a pleasure to recognize their achievements by including their names on the next page.

Finally, I would like to thank all the Wiley people who did such a marvelous job with the production of the book. In chronological order of service they are David Kaplan, Beatrice Shube (Field Marshall), Robert Polhemus, Robert Golden, Kenneth McLeod, Rose Ann Campise, and John Russell.

EDGAR M. PALMER

East Lansing, Michigan
March 1985

RANDOM GRAPH THEORISTS

The difference is people

IBM MOTTO

S. M. Selkow, L. Kučera, A. R. Bloemena, E. Shamir, J. I. Hall, R. T. Smythe, N. Sauer, P. E. O'Neill, H. L. Frisch, L. G. Valiant, A. Békéssy, J. B. Kalbfleisch, J. M. Hammersley, J. W. Moon, V. A. Perepelica, W. Kordecki, Z. Palka, K. Schürger, R. L. Graham, V. L. Klee, L. Katz, G. B. Smirnova, A. M. Murray, M. Dondajewski, I. N. Kovalenko, C. J. H. McDiarmid, G. S. Plesnevich, G. Cornuejols, P. Camion, J. Komlós, L. Babai, J. Spencer, I. Frisch, P. M. D. Gray, V. Chvátal, R. Van Slyke, W. F. Penney, V. P. Chistyakov, S. A. Burr, J. D. Isaacson, R. S. Wilkov, A. Kershenbaum, V. N. Lyamin, D. G. Corneil, I. C. Rényi, J. H. Redfield, O. S. Krakovskaia, V. E. Stepanov, E. N. Gilbert, N. G. deBruijn, W. Feller, B. A. Sevast'yanov, E. I. Litvak, J. Wierman, A. G. Thomason, P. Holgate, M. Ajtai, C. R. Darwin, L. Pósa, R. W. Robinson, R. J. Wilson, E. H. Gimadi, B. I. Selivanov, G. N. Bagaev, J. V. Schulz, F. Proschan, E. A. Bender, J. C. Ogilvie, J. E. Cohen, E. F. Moor, R. E. Barlow, G. Szekeres, Yu. L. Pavlov, P. J. Cameron, R. C. Read, G. I. Ivchenko, Y. Fu, K.-J. Thürlings, V. M. Chelnokov, A. D. Koršunov, C. W. Marshall, C. C. Rousseau, A. Rapoport, T. L. Austin, W. Gaul, E. K. Lloyd, P. Békéssy, S. O. Rice, G. E. Uhlenbeck, N. D. Angluin, G. G. Killough, H. Hadwiger, K. Shütte. P. L. Chebyshev, S. Ulam, F. Ramsey, F. R. K. Chung, C. M. Grinstead, B. L. Rothschild, J. W. Essam, F. Harary, P. A. Catlin, V. N. Sachkov, D. S. Johnson, A. Ruciński, J. Szymański,

M. F. Capobianco, E. Upfal, N. C. Wormald, M. V. Lomonosov, M. R. Garey,
V. Palmer, M. S. Saparov, V. P. Polessky, A. Grusho, M. A. Zaitsev, I. V.
Medvedev, Yu. D. Burtin, J. Riordan, S. Fajtlowicz, W. F. de la Véga, B. R.
Heap, O. Frank, E. R. Canfield, R. H. Schelp, C. King, E. Szemerédi, A. A.
Kalnin'sh, A. Meir, R. J. Riddell, E. M. Palmer, A. K. Kel'mans, L. Moser,
A. J. Schwenk, Ya. M. Barzdin, H. S. Na, A. Cayley, G. A. Margulis, B. Bollobás,
Ch. Jordan, D. E. Knuth, A. J. Tolcan, A. Rényi, A. Blass, G. Pólya, D. W.
Matula, R. J. Faudree, M. Karoński, S. S. Yan, D. G. Larman, G. R.
Grimmett, G. Marble, C. K. Bailey, C. E. Bonferroni, R. Ling, N. A. Young,
A. D. Barbour, B. A. Trahtenbrot, H. Frank, L. J. Hubert, V. V. Epihin, V. P.
Kozyrev, E. M. Wright, C. E. Shannon, T. Odda, J. W. Kennedy, L. Lovász, P.
Erdös, I. Palásti, R. E. Fagen, Le Cong Thanh, V. F. Kolchin, A. W. A.
Shogan, L. Rabinowitz, L. E. Clarke, R. Otter, T. I. Fenner, B. D. McKay,
A. M. Frieze, H. S. Wilf, J. D. Dixon, A. Nijenhuis, W. Oberschelp, N. L. Biggs,
H. Prüfer, G. Tinhofer, I. M. H. Etherington, W. W. Kuhn, S. M. Ross, M. J.
Byrne, C. Jordan, V. Rödl, E. Toman, T. Mueller, I. Tomescu, A. D. Glukhov,
M. Sulyok, F. Juhasz, J. I. Naus, B. Pittel, D. Oberly, D. P. Sumner, M. M.
Matthews, L. Guy, P. K. Stockmeyer, V. G. Vizing, V. Müller, R. J. Lipton,
R. M. Karp, M. Las Vernas, E. L. Lawler, D. W. Walkup, R. L. Hemminger,
H. Whitney, J. A. Bondy, R. L. Brooks, W. T. Tutte, G. A. Dirac, K. Menger,
K. Kuratowski, J. P. C. Petersen, O. Ore, L. Euler.

Exercise 0.0.0

If these names were rearranged in alphabetical order, how many could be
expected to maintain the same position?

CONTENTS

A man in the house is worth two in the street.

MAE WEST

Belle of the Nineties

SYMBOLS

*Business was so bad the other night the orchestra was
playing "Tea for One."*

HENNY YOUNGMAN

A	probability Model A
$\mathscr{A}$	a set of labeled graphs of order n
$\bar{\mathscr{A}}$	complementary set
B	probability Model B
$\mathscr{B}$	set of labeled graphs of order n with no isolated vertices
ccl	contraction clique number
cl	clique number
$c_{n,q}$	number of unlabeled connected graphs of order n and size q
C	probability Model C
C_n	number of labeled connected graphs of order n
$C_{n,q}$	number of labeled connected graphs of order n and size q
$\mathscr{C}$	set of labeled connected graphs of order n
deg v	degree of v
$d(x)$	generating function for vertices of degree 1 in trees
$D(x)$	generating function for vertices of degree 1 in rooted trees
$E = E(G)$	edge set
$E(X)$	expectation

$E_r(X)$	rth factorial moment
$f(x)$	generating function for fixed vertices (of degree r) in trees
$F(x)$	generating function for fixed vertices (of degree r) in rooted trees
F	finite permutation group
F_i	ith conjugacy class of F
g	girth of a graph
g_n	number of unlabeled graphs of order n
$g_{n,q}$	number of unlabeled graphs of order n and size q
G	a labeled graph
G_n	number of labeled graphs of order n
$G_{n,q}$	number of labeled graphs of order n and size q
$G_n^{(r)}$	number of labeled r-regular graphs of order n
$\mathcal{G}_n^{(r)}$	set of labeled r-regular graphs of order n
H	a subgraph
$H_{n,q}$	the number of labeled hamiltonian graphs of order n and size q
(j)	partition of n
K_m	complete graph of order m
$K_{m,n}$	complete bipartite graph
$l(G)$	number of ways to label G
n	number of vertices or order of a graph
$(n)_k$	falling factorial
$n^*(T)$	number of ways to root T at a vertex
$N(F)$	number of orbits of F
o	the little oh function
O	the BIG OH function
$\mathcal{O}$	orbit of a permutation group
$O_s(x)$	generating function for vertices of degree r and orbit size s in rooted trees
p	probability of an edge
$P(\mathcal{A})$	probability of set $\mathcal{A}$
$P(X \geq 1)$	probability of event $x \geq 1$, etc.
q	number of edges or size of a graph
$q^*(T)$	number of ways to root T at an edge

$r(k, l)$	Ramsey number
r_n	number of rooted trees of order n
$s(G)$	number of symmetries of G or the order of its automorphism group
$s(T)$	number of symmetry edges of T
$s(r, j)$	Stirling numbers of the first kind
$S(r, j)$	Stirling numbers of the second kind
S_r	rth binomial moment or symmetric group
t_n	number of unlabeled trees of order n
$t_n^{(1)}$	number of trees with a single centroid
$t_n^{(2)}$	number of trees with a double centroid
$t(x)$	generating function for trees
T	for tree or tournament
T_n	number of labeled trees of order n
$T^{(r)}(x)$	generating function for trees of root degree r
$T(x)$	generating function for rooted trees
u, v, w	vertices of a graph
$V = V(G)$	vertex set
$V(X)$	variance
X, Y	random variables
$Z(S_r)$	cycle index of symmetric group of degree r
β	independence number
$\Gamma(G)$	automorphism group of G
δ	minimum degree
Δ	maximum degree
κ	connectivity
λ	edge-connectivity
μ	mean
ρ	radius of convergence of $T(x)$
σ	variance
χ	chromatic number
ω_n	function of n that approaches infinity arbitrarily slowly
Ω	sample space

1

INTRODUCTION

My brain is open.

PAUL ERDÖS

Graphical enumeration is the study of special integer sequences. One wants to know how to compute effectively the elements of the sequence and determine their behavior asymptotically. The subject was pioneered by Cayley, Redfield, Pólya, Otter and Wright and has close ties to the theory of evolution of random graphs founded by Erdös and Rényi. Here is a brief review of some important points.

1.1. A LITTLE HISTORY

Let $V = \{v_1, \ldots, v_n\}$ be a finite set of elements called *vertices*, and let E be a subset of the collection of unordered pairs of vertices. The elements of E are called *edges*. A *graph G* consists of an ordered pair (V, E) of sets of vertices and edges, respectively. The number n of vertices is called the *order* of G, while its *size* is the number q of edges. In a labeled graph the vertices have fixed identities, and we sometimes say that the vertex v_i has label i.

One of the fundamental distributions that has received much attention is formed by the graphs of order n classified by size. This distribution for unlabeled graphs is illustrated in Figure 1.1.1.

1

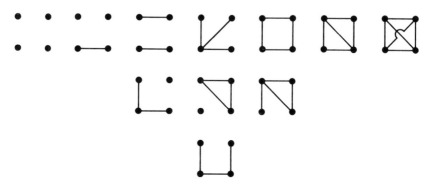

FIGURE 1.1.1.   The unlabeled graphs of order 4.

The following formula for the number, $l(G)$, of ways to label a given unlabeled graph G is regarded as enumerative folklore. It depends on the number of symmetries of G, that is, the order of the automorphism group $\Gamma(G)$:

$$(1.1.1) \qquad\qquad l(G) = \frac{n!}{|\Gamma(G)|}.$$

The proof is made by first labeling the vertices of G in all $n!$ ways. But these labelings will be equivalent whenever they differ by an automorphism. So the $n!$ labelings are partitioned into $l(G)$ classes, each with $|\Gamma(G)|$ equivalent labelings.

Naturally this type of result is useful for all sorts of finite structures, not just for graphs. Note also that the number of ways to label G is just the same as the number of ways to label the complement $\overline{G}$.

We denote the number of labeled graphs of order n by G_n and the number of size q by $G_{n,q}$. The next two formulas are elementary:

$$(1.1.2) \qquad\qquad G_n = 2^{\binom{n}{2}}$$

$$(1.1.3) \qquad\qquad G_{n,q} = \left(\binom{\binom{n}{2}}{q} \right).$$

On the other hand, the corresponding formulas for unlabeled graphs are not elementary at all. We use g_n for the number of unlabeled graphs of order n and $g_{n,q}$ for those of size q. Thus, for example, we see from Figure 1.1.1 that $g_4 = 11$ and $g_{4,3} = 3$.

J. H. Redfield [R27] was the first to find an efficient method for calculating these, and he used his superposition theorem to determine g_n for $n \le 7$.

Unfortunately, his paper was completely overlooked until it was discovered by Harary around 1959 (see [H60]). A biographical sketch [Ll84] and commentaries ([HR84], [HaPR84]) on the work of this remarkable scholar can be found in the issue of the *Journal of Graph Theory* dedicated to his memory.

Pólya used his multipurpose enumeration theorem [P37] to find the number of graphs (see [H55a], where g_n for $n \leq 7$ is also computed). His formulas are stated in Section 6.1, where they are needed for the generation of random unlabeled graphs. They were used by Clarence King of the University of the West Indies to compute the number of graphs of order $n \leq 24$. King's results appeared in [HP73].

Pólya also determined asymptotic formulas for g_n and $g_{n,q}$. In particular he observed the following relations between labeled and unlabeled graphs:

$$(1.1.4) \qquad g_n \sim \frac{2^{\binom{n}{2}}}{n!}$$

and

$$(1.1.5) \qquad g_{n,q} \sim \frac{\binom{\binom{n}{2}}{q}}{n!}$$

provided q is near the center of its range.

A brief sketch of Oberschelp's proof of (1.1.4) is given in Section 6.1. All the details for the verification of (1.1.4) and (1.1.5) appear in Chapter 9 of [HP73].

E. M. Wright [W70b] was even able to find a necessary and sufficient condition for (1.1.5) to hold; namely,

$$(1.1.6) \qquad \left[\frac{2}{n} \min\left\{ q, \binom{n}{2} - q \right\} - \log n \right] \to +\infty.$$

The proof of the sufficiency of this condition is quite complicated. But the necessity is rather easy to explain once one is familiar with a few basic ideas from the theory of random graphs developed in this monograph (see exercise 4.3.2). The importance of the theorem of Wright will be seen in Section 4.3, where it is used to translate information about labeled graphs into results for unlabeled graphs.

Wright was also interested in proving the unimodality of the distribution of unlabeled graphs. He succeeded for n sufficiently large [W74d], but later the result for all n was shown to follow from an application of a very general theorem of Thürlings [Th78].

Some of the earliest enumeration theorems in graph theory are concerned with the graphs of small size in our distribution. These are the work of the prolific Victorian mathematician A. Cayley, who tackled the problem of determining the number of trees of order n. Some of his papers on this subject are reprinted in the excellent history book of Biggs, Lloyd and Wilson [BiLW76].

A tree is a connected graph that has no cycles. Evidently Cayley first derived recurrence relations for the number t_n of unlabeled trees of order n from which he was able to calculate the numbers for $n \le 13$. His formulas for rooted and unrooted trees are given in Section 6.1. Later he found the beautiful formula for the number T_n of labeled trees of order n.

Cayley's Formula

$$(1.1.7) \qquad\qquad T_n = n^{n-2}.$$

This discovery is unavoidable if one just tabulates the numbers of labeled trees of order $n \le 5$. On the other hand, there are many ways to prove (1.1.7), and none of them is completely obvious. J. W. Moon [M67] shed light on this matter by collecting ten distinct proofs. Professor Erdös has said that there is a Big Book that contains the very best proofs of all theorems but only very rarely are we allowed a glimpse of one of its pages. No doubt Prüfer's proof [Pr18] of Cayley's formula is one from the Book; it is also found in almost every graph theory text. His technique is discussed at some length in Section 6.1. Many other interesting results are in the standard reference for labeled trees, J. W. Moon's book [M70], whose bibliography lists over 160 titles.

Both Otter and Pólya made significant contributions to the enumeration of unlabeled trees. Pólya [P37] used generating functions to count trees whose vertices have degree 1 or 4 and analytic function theory to determine asymptotic behavior. Otter [O48] built on and improved the work of Cayley and Pólya. In particular, he found the asymptotic behavior of the number t_n of unlabeled trees to be given by

$$(1.1.8) \qquad\qquad t_n \sim \frac{.5349\ldots}{n^{5/2}\rho^n}$$

where $\rho = .3383219\ldots$ is the radius of convergence of the generating function $t(x) = \Sigma t_n x^n$. Some of these results will be used in Section 6.3 when we study the distribution of the vertices in trees by degree and orbit size.

Although much can be learned about the distribution of Figure 1.1.1 by applying enumeration techniques, its nature may be more deeply revealed by the theory of graphical evolution founded by Erdös and Rényi [ErR60]. In the following chapters we shall discuss this theory in some detail, concentrating on the labeled case.

2

PROBABILITY MODELS
FOR GRAPHS

*It's a chancy job but it makes a man watchful and
sometimes a little lonely.*

MATT DILLON
U.S. MARSHALL

We introduce three probability models for graphs and show how to use some simple concepts of probability theory to answer a number of questions about the asymptotic behavior of the number of labeled graphs with special properties such as connectivity. The reader may find Feller's introductory text [Fe57] on probability theory helpful for concepts not covered here or in the Appendix.

2.1. MODELS A, B AND C

Suppose we wish to estimate the number of connected, labeled graphs of large order n. A sensible approach to this problem involves the study of the proportion of labeled graphs that are connected. Let C_n denote the number of connected, labeled graphs of order n while $C_{n,q}$ denotes the number of these with size q. To investigate the question raised above, we should consider both of the ratios C_n/G_n and $C_{n,q}/G_{n,q}$. In light of formulas (1.1.2) and (1.1.3), both

5

ratios could be analyzed with the aid of formulas for C_n and $C_{n,q}$. Nice recurrence relations do exist for these functions. Perhaps the easiest to derive for C_n is the following simple expression (see [Ri51], [RiU53] or [G56]):

$$(2.1.1) \qquad\qquad nG_n = \sum_{k=1}^{n} \binom{n}{k} k C_k G_{n-k}$$

where $G_0 = 1$. This formula can be verified by counting rooted labeled graphs in two different ways. A labeled graph of order n can be rooted at any of its n vertices. Hence nG_n is the number of these. On the other hand, the root must occur in a component of order $k = 1$ to n. The binomial coefficient $\binom{n}{k}$ is the number of ways to select k labels for the vertices of the component containing the root. The product kC_k is the number of rooted components with these labels, and G_{n-k} allows any graph at all to occupy the remaining $n - k$ vertices. Thus the sum on the right side of (2.1.1) also counts rooted labeled graphs of order n.

In a few minutes with a pocket calculator, formula (2.1.1) can be used to determine C_n/G_n for $n \leq 8$. For example, $C_8/G_8 = .937\ldots$, and in fact it is easy to see that the sequence C_n/G_n is strictly increasing with limit 1 (see exercise 2.1.5).

There are also many formulas for $C_{n,q}$, but the recurrence derived by Wormald [Wo78c] is probably the most convenient for purposes of calculation:

$$(2.1.2)$$

$$(q + 1)C_{n,q+1} = \left[\binom{n}{2} - q\right]C_{n,q} + \frac{1}{2}\sum_{i=1}^{n-1}\sum_{j=1}^{q} \binom{n}{i} i(n - i)C_{i,j}C_{n-i,q-j}$$

for $n > 1$ and all q, with $C_{1,0} = 1$ and $C_{n,0} = 0$ for $n > 1$. Note that the left side of (2.1.2) is the number of edge-rooted connected graphs. The first term on the right side counts those for which deletion of the root edge leaves a connected graph. The double sum counts the ones whose root edge is a bridge.

Values of $C_{n,q}$ for $n \leq 8$ are available in [Wo78c]. No doubt certain aspects of the asymptotic behavior of $C_{n,q}$ could be determined from (2.1.2), but there are other recurrence relations more suitable for such analysis. Wright [W70a] has used such an approach to investigate successfully the range of q for which $C_{n,q} \sim G_{n,q}$ (see exercise 2.1.9). But we shall be concerned with a broader range for q, and it seems that such a study is most natural in the context of probability theory. Therefore we introduce the first of three probability models that will be used in this text.

Model A: For each positive integer n and each number p with $0 < p < 1$, the sample space Ω consists of all labeled graphs G of order n. If G has q

edges, the probability of G is given by

(2.1.3)
$$P(G) = p^q (1 - p)^{\binom{n}{2} - q}.$$

The fact that $P(\Omega) = 1$ follows immediately from the binomial theorem.

Here is a useful way of thinking about this model. Consider n labeled vertices. Therefore there are $\binom{n}{2}$ slots available for edges. Arrange the slots in some linear order, and in that order, one slot at a time, independently, introduce an edge with probability p. When the process terminates, a "random (labeled) graph" will have been generated. Sometimes the number p is called the "probability of an edge."

Note that if $p = 1/2$, then each graph is assigned the same probability, namely, $1/G_n$. Therefore in this special case the probability of the set of connected graphs is C_n/G_n, exactly the ratio considered earlier.

Observe also from (2.1.3) that if $p < 1/2$, the more edges a graph has, the lower is its probability. Furthermore, the smaller p is, the less likely is the random graph to be connected.

Nevertheless, in one of the first papers on random graphs [G59], Gilbert showed that when p is fixed, the probability of connectivity approaches 1 as its limiting value when $n \to \infty$. Specifically, for each n let P_n denote the probability of the set of connected graphs of order n. Then by definition

(2.1.4)
$$P_n = \sum_{q=n-1}^{\binom{n}{2}} C_{n,q} p^q (1 - p)^{\binom{n}{2} - q},$$

but the behavior of P_n is much easier to determine from the recurrence relation found by Gilbert:

(2.1.5)
$$1 = \sum_{k=1}^{n} \binom{n-1}{k-1} P_k (1 - p)^{k(n-k)},$$

which is just a slight generalization of our earlier formula (2.1.1). It is easy to use (2.1.5) to show that

(2.1.6)
$$P_n = 1 - O\left(n(1 - p)^{n/2}\right).$$

To see this, just solve equation (2.1.5) for P_n and find

(2.1.7)
$$P_n = 1 - F_n,$$

where

$$(2.1.8) \qquad F_n = \sum_{k=1}^{n-1} \binom{n-1}{k-1} P_k (1-p)^{k(n-k)}.$$

Since we are expecting P_k to be close to 1 for large k and since $P_k \le 1$, it is reasonable to eliminate it from the right side of (2.1.8) to obtain a good upper bound for F_n. This bound will be symmetric in k, and so we have

$$(2.1.9) \qquad F_n \le 2 \sum_{k=1}^{\lfloor n/2 \rfloor} \binom{n-1}{k-1} (1-p)^{k(n-k)}.$$

Now we can use the crude bounds $\binom{n-1}{k-1} \le n^k$ and $n - k \ge n/2$ to establish

$$(2.1.10) \qquad F_n \le 2 \sum_{k=1}^{\lfloor n/2 \rfloor} \left[n(1-p)^{n/2} \right]^k.$$

Since p is fixed,

$$(2.1.11) \qquad n(1-p)^{n/2} = o(1).$$

Therefore for n sufficiently large, the right side of (2.1.10) is the partial sum of a convergent geometric series. Since this sum starts at $k = 1$,

$$(2.1.12) \qquad F_n = O\!\left(n(1-p)^{n/2} \right),$$

which establishes (2.1.6). Thus for fixed p, $\lim_{n \to \infty} P_n = 1$ (we often write $P_n \to 1$ for short).

If $\mathscr{A}$ is any set of labeled graphs of order n with property Q such that $\lim_{n \to \infty} P(\mathscr{A}) = 1$, we say that "almost all graphs have property Q." Again note that we often write $P(\mathscr{A}) \to 1$. Thus in Model A with p fixed, formula (2.1.6) implies that almost all graphs are connected.

Recall that we also wish to study the ratio $C_{n,q}/G_{n,q}$. For this reason we introduce our second probability model.

Model B: For each positive integer n and integer q between 0 and $\binom{n}{2}$, the sample space Ω consists of all labeled graphs G of order n and size q. The probability of G is given by

$$(2.1.13) \qquad P(G) = \left(\binom{\binom{n}{2}}{q} \right)^{-1}.$$

Therefore, in Model B, the probability of connectivity is $C_{n,q}/G_{n,q}$. Note that the simple inequality of exercise 2.1.6 shows that for a given n this probability increases from 0 to 1. This suggests that to determine when almost all graphs are connected, we need to find a very special function $q = q(n)$ that is in some sense the smallest one that will still permit the probability of connectivity to have limit 1 as $n \to \infty$.

This seems to be a much more difficult question than the one we encountered with Model A. Nevertheless, the two models are very closely related, as indicated by exercise 2.1.9, where there is an opportunity to appreciate the techniques used in work with Model B. We shall return to the important problem raised above when we are in a better position to take advantage of the relationship between the two models.*

Another model was suggested by Erdös and Rényi [ErR60], but only recently has it received much attention. Continuing under the influence of Detroit and the automobile industry, we will call it Model C.

Model C: For positive integers n and r with $1 \le r \le n - 1$, the sample space Ω consists of all labeled digraphs of order n in which each vertex has outdegree r. Each digraph D is given the probability

(2.1.14) $$P(D) = \binom{n-1}{r}^{-n}.$$

For a given vertex v, there are $\binom{n-1}{r}$ possible choices for the vertices adjacent from v. On applying the multiplication principle over all n vertices, we find that there are $\binom{n-1}{r}^{n}$ digraphs in the sample space. Hence $P(\Omega) = 1$.

In our studies of Model C we will always assume that r is fixed. Occasionally we will ignore the orientation of the arcs of the digraphs in the sample space to obtain a probability model for graphs. For example, in Model C the probability of connectivity for graphs is the same as the probability of weak connectivity for digraphs.

Exercises 2.1

1. In Model A with $n > 10^6$, what is the sum of the probabilities of all graphs in which the two vertices with labels 1984 and 1776 are adjacent?

2. Find the probability that a graph of order 4 is connected in Model A with $p = 3/5$.

3. What is the probability that a graph of order 4 has isolated vertices in Model A with $p = 2/5$?

---

*Since we have offered more than one model, this text may be legal in Texas, where the State Board of Education requires presentation of alternative explanations of evolutionary processes.

4. Find the probability that a graph of order 4 and size 3 is connected in Model B.

5. Prove that the ratio C_n/G_n is strictly increasing for $n \geq 3$.

6. When is it true that $C_{n,q}/G_{n,q} \leq C_{n,q+1}/G_{n,q+1}$?

7. Let R_n be the number of labeled tournaments of order n in which v_1 is a source, that is, every vertex is reachable from v_1. Find a recurrence relation for R_n and show that $R_n/2^{\binom{n}{2}} \to 1$.

8. Work out a recurrence relation for the number W_n of weakly connected digraphs of order n in the sample space of Model C.

9. Reformulate equation (2.1.1) for graphs of order n and size q. Now fix p, $0 < p < 1$, and in Model B with $q = \left\lfloor p\binom{n}{2} \right\rfloor$ prove that almost all graphs are connected.

2.2. EXPECTATION

The study of the limiting value of probabilities for graphs can be substantially facilitated by the use of the expectation of random variables. In the previous section we showed that when p is fixed, almost all graphs are connected in Model A. The same is true in Model B for $q = \left\lfloor p\binom{n}{2} \right\rfloor$. This result can also be proved, as asked for in exercise 2.1.9, by using an explicit formula for the number of connected graphs. There is a probabilistic approach, however, that leads to the same conclusion without the necessity (or the benefits) of enumerating connected graphs.

For each graph G in the sample space, let $X(G) = 1$ or 0 according as G is disconnected or not. Then $E(X)$, the expected number of disconnected graphs, is the sum of the probabilities of all disconnected graphs. Therefore in Model A,

$$(2.2.1) \qquad E(X) \leq \sum_{k=1}^{n-1} \binom{n}{k}(1-p)^{k(n-k)}$$

because the right side of (2.2.1) is the sum of the probabilities of all graphs whose vertices have been partitioned into two ordered nonempty sets with no edges joining vertices of different sets. The inequality is necessary because every disconnected graph can be so partitioned in at least two ways, perhaps more.

But a simple consequence of Markov's inequality (see Proposition 5.2 of Appendix V) that we will use over and over again shows that $P(X \geq 1) \leq$

$E(X)$. Therefore, to show that almost all graphs are connected we need only show that $E(X) \to 0$; that is, if you expect no disconnected graphs, then there are almost none. Therefore almost all graphs are connected in Model A with p fixed if the right side of (2.2.1) goes to zero. But this can be done with exactly the same few steps used to establish (2.1.6).

The same conclusion can be reached even more easily by a better choice for the random variable. For each graph G in the sample space of Model A, now let the random variable $X(G)$ be the number of unordered pairs $\{u, v\}$ of vertices of G such that no other vertex of G is adjacent to both u and v. Think of these pairs as "bad," whereas a pair of vertices is "good" if it is joined by a path of length 2. Of course, if a graph has no bad pairs of vertices, it must be connected. Putting this another way, if $\mathscr{C}$ is the set of connected labeled graphs of order n, then $P(X = 0) \le P(\mathscr{C})$ or

$$(2.2.2) \qquad\qquad P(\overline{\mathscr{C}}) \le P(X \ge 1).$$

Therefore, to show that almost all graphs are connected, we need only find $E(X)$ and show that $E(X) \to 0$; that is, if you expect no bad pairs, then almost no graph has any.

To illustrate an approach that will be useful in other cases, we will determine $E(X)$ in more detail than necessary. First we list the pairs of vertices from 1 to $t = \binom{n}{2}$ and write the random variable X as a sum

$$(2.2.3) \qquad\qquad X = X_1 + \cdots + X_t,$$

where $X_i(G) = 1$ or 0 according as the ith pair of vertices of G is bad or good, respectively. Then, by the linearity of expectation,

$$(2.2.4) \qquad\qquad E(X) = E(X_1) + \cdots + E(X_t).$$

But in this particular case we have $E(X_1) = \cdots = E(X_t)$, and therefore

$$(2.2.5) \qquad\qquad E(X) = \binom{n}{2} E(X_1).$$

By definition, $E(X_1)$ is the sum of the probabilities of all graphs for which no vertex w is adjacent to both u and v. The probability that a particular vertex is not adjacent to both u and v is $1 - p^2$. This probability is independent of the probability that any other vertex is not adjacent to both u and v. Therefore the probability that none of the other $n - 2$ vertices are adjacent to both u and v is obtained by multiplication, and we have

$$(2.2.6) \qquad\qquad E(X_1) = \left(1 - p^2\right)^{n-2},$$

and therefore

$$(2.2.7) \qquad E(X) = \binom{n}{2}(1 - p^2)^{n-2}.$$

Obviously, $E(X) \to 0$ for p fixed, and so almost all graphs have no bad pairs of vertices. Thus with very little effort we have shown that almost all graphs are connected.

Here is the expectation for the same random variable but in Model B:

$$(2.2.8) \quad E(X) = \binom{n}{2}\sum_{k=0}^{n-2}\binom{n-2}{k}\left(\binom{\binom{n}{2} - (n-2) - k}{q - k}\right)\left(\binom{\binom{n}{2}}{q}\right)^{-1}.$$

In this formula, think of k as the number of vertices ($\neq u, v$) that are adjacent to u. The binomial coefficient $\binom{n-2}{k}$ selects the k slots for these edges, and the next coefficient selects the slots for the remaining $q - k$ edges so that no vertex is adjacent to both u and v.

If we take $q = \left\lfloor p\binom{n}{2} \right\rfloor$ for fixed p, $0 < p < 1$, then it can be shown that $E(X)$ as given by (2.2.8) also has limit 0 as $n \to \infty$. However, it is evidently more difficult to deal with $E(X)$ in Model B than in Model A.

This approach for demonstrating connectivity in Models A and B also works in Model C. For each digraph D of order n in the sample space of Model C, the random variable $X(D)$ counts the unordered pairs $\{u, v\}$ of vertices such that no other vertex is adjacent to or from both u and v. That is, u and v are not joined by a semipath of length 2. These are the bad pairs of vertices, and the expected number of them is

$$(2.2.9) \qquad E(X) = \binom{n}{2}\binom{n-2}{r, r, n-2-2r}\binom{n-2}{r}^{2r}$$
$$\times \left[2\binom{n-3}{r-1} + \binom{n-3}{r}\right]^{n-2r-2}\binom{n-1}{r}^{-n}.$$

The binomial coefficient $\binom{n}{2}$ in (2.2.9) is the number of ways to select vertices u and v. The multinomial coefficient is the number of ways to select disjoint sets S and T of r vertices adjacent from u and r vertices adjacent from v. If vertex w belongs to S, it cannot be adjacent to v, so there are $n - 2$ vertices from which the r vertices adjacent from w can be selected. Similar reasoning if w belongs to T leads to the third factor $\binom{n-2}{r}^{2r}$ of the right side of (2.2.9). The next factor is contributed by the $n - 2r - 2$ remaining vertices. If one of these is adjacent to u, it cannot be adjacent to v, and vice versa. This accounts for the term $2\binom{n-3}{r-1}$, and the other term $\binom{n-3}{r}$ allows for the option of a vertex being adjacent to neither u nor v.

The following asymptotic estimate of $E(X)$ in (2.2.9) can be established using no more than the definition of the binomial coefficients:

(2.2.10) $E(X) \sim \frac{1}{2}(r!)^{2r-1} n^{2+2r-2r^2}$.

Thus for fixed $r \geq 2$, $E(X) \rightarrow 0$, and we have a connectivity theorem for Model C.

Theorem 2.2.1. In Model C with fixed $r \geq 2$, almost all digraphs are weakly connected.

Note that sometimes we ignore the orientation of the arcs of the digraphs in Model C, and then we say that almost all graphs in Model C are connected. This theorem was proved in more generality by Fenner and Frieze [FeF82], and we will discuss their stronger result in Chapter 5.

Exercises 2.2

1. A component of order 2 in a graph is sometimes called an isolated edge. Find the expected number of isolated edges in Model A and in Model B.

   Let $X(G)$ be the number of edges of G that are not in triangles (complete subgraphs of order 3).

2. Find $E(X)$ in Model A.

3. Find $E(X)$ in Model B.

   Let $X(G)$ be the number of vertices of G that are not in triangles.

4. Find $E(X)$ in Model A.

5. Find $E(X)$ in Model B.

2.3. PROPERTIES OF ALMOST ALL GRAPHS

The basic idea concerning bad pairs of vertices was generalized by Blass and Harary [BlH79] for the purpose of studying other properties of almost all graphs.

Theorem 2.3.1. Integers $k \geq 0$ and $l \geq 0$ are fixed, and so is the probability p of an edge. Then, in Model A, almost every graph has the property that for any disjoint sets of vertices S and T of order $\leq k$ and $\leq l$, respectively, some vertex is adjacent to each vertex of S but not to any vertices of T.

Proof. A pair of sets S and T is "bad" if no vertex is adjacent to each vertex of S but not to any of T. Let $X(G)$ be the number of bad pairs of sets S and T of order exactly k and l, respectively. Then the expectation of X is

$$(2.3.1) \qquad E(X) = \binom{n}{k, l, n - k - l}\left[1 - p^k(1 - p)^l\right]^{n-k-l}.$$

Note that when $k = 2$ and $l = 0$, this formula becomes (2.2.7). And $E(X) \to 0$ here also. Thus for fixed k, l and p we have $E(X) \to 0$, and so once again there are no bad pairs of sets of this order in almost every graph. Obviously, this is sufficient to prove the result for sets of smaller order as well. □

Quite a number of properties of almost all graphs can now be established by successive applications of this theorem. The following list is valid for fixed k and Model A with p fixed.

(i)   Almost all graphs have diameter 2.

(ii)   Almost all graphs are k-connected.

(iii) Almost all graphs contain a given subgraph of order k as an induced subgraph.

(iv)   Almost all graphs are nonplanar.

(v)   Almost all graphs are locally connected, that is, the deleted neighborhood of *every* vertex is connected.

Observe that Theorem 2.3.1. is used to show that almost all graphs have diameter 2, and as a consequence almost all graphs are connected. This would lead one to suspect that there may be interesting properties held by almost all graphs that cannot be proved simply by applying this theorem. In fact, we will see later that almost all graphs are hamiltonian but a graph without certain bad sets is not necessarily hamiltonian. Hence much more than Theorem 2.3.1 would be required to show that almost all graphs are hamiltonian. See [BlH79] and the related reference [BoP77] for a detailed discussion of the scope of this theorem.

On the other hand, the underlying idea of the theorem can be of use in slightly different contexts. For example, there is a question known in the literature as Shütte's problem, which asks if there exists a (round-robin) tournament with the following property for fixed k:

For every k-set S of players, there is another player who dominates everyone in S.

It is a simple matter to set up a probability model to handle this question. The sample space consists of all $2^{\binom{n}{2}}$ labeled tournaments of order n. One player dominates another with probability $1/2$ (compare Model A!), and so each tournament has the same probability. A k-set of players is "bad" if no other player dominates its players. If $X(T)$ is the number of bad k-sets in the tournament T, then

(2.3.2) $$E(X) = \binom{n}{k}\left[1 - (1/2)^k\right]^{n-k}$$

and as usual $E(X) \to 0$. Therefore, not only do such tournaments exist, but for a given fixed k, when n is large, the overwhelming majority of tournaments will have this property.

Nevertheless, it is not so easy to construct them, even when $k = 2$. As soon as the right side of (2.3.2) is < 1, then we know such a tournament exists. For example, it is < 1 for $k = 2$ and $n = 22$. However, the smallest such tournament (see [GrS71]) has only seven vertices!

Possible generalizations of the theorem are suggested by exercise 2.3.5 at the end of this section.

We can also strengthen the weak property (ii) for connectivity by determining how fast $m = m(n)$ can grow so that in Model A almost all graphs are still m-connected. For each graph G in the sample space, let the random variable $X(G)$ be the number of unordered pairs $\{u, v\}$ of vertices of G such that fewer than m other vertices are adjacent to both u and v. These are the "bad" pairs of vertices. A graph with no bad pairs is good; that is, it is m-connected because every pair of vertices are joined by at least m disjoint paths. Here is the expectation of X:

(2.3.3) $$E(X) = \binom{n}{2} \sum_{k=0}^{m-1} \binom{n-2}{k}(p^2)^k(1 - p^2)^{n-2-k}.$$

This formula generalizes (2.2.7), which is just (2.3.3) with $m = 1$. As before, $\binom{n}{2}$ is the number of ways to select two vertices u and v. The binomial coefficient $\binom{n-2}{k}$ is the number of selections of k other vertices to be adjacent to both u and v. For each of these k vertices the probability is p^2 that it is adjacent to both u and v, and for the other $n - 2 - k$ vertices the complementary probability $1 - p^2$ is used.

Formula (4.3) of Proposition 4.1 in Appendix IV provides an upper bound for the lower tail of the binomial distribution, and on applying it to (2.3.3) we find that if

(2.3.4) $$m - 1 \le p^2(n - 2),$$

then

(2.3.5) $$E(X) \le \binom{n}{2}\binom{n-2}{m-1}(p^2)^{m-1}(1 - p^2)^{n-1-m}F(n, m),$$

where

(2.3.6)
$$F(n, m) = \frac{p^2(n - m)}{p^2(n - 1) - (m - 1)}.$$

Therefore if $0 < \varepsilon < 1$ and

(2.3.7)
$$\frac{m}{n} \leq \varepsilon p^2,$$

we can say

(2.3.8)
$$E(X) = O(1) n^2 \binom{n - 2}{m - 1} p^{2m} (1 - p^2)^{n - m}$$
$$= O(1) \frac{n^{m+1} p^{2m} (1 - p^2)^{n - m}}{(m - 1)!}.$$

On taking the logarithm of each side of (2.3.8), in just a few steps we find that $\log E(X) \to -\infty$ provided

(2.3.9)
$$m = o\left(\frac{n}{\log n}\right).$$

This useful result is summarized next.

Theorem 2.3.2. In Model A with the probability p of an edge fixed and $m = o(n/\log n)$, almost every graph has at least m paths of length 2 joining every pair of vertices.

This means, for example, that if $m = n^{1-\varepsilon}$ where $\varepsilon > 0$ is very small, almost every graph is m-connected! But later we will do even better than this (see exercise 5.1.5).

Exercises 2.3

1. Do almost all graphs have the property that every edge is in a triangle? What about every vertex?

2. Define a digraph of order 7 with vertices $v_0, \ldots, v_6$ and v_i adjacent to v_j if $j - i$ is a nonzero square in the field of order 7. Check that this digraph is, in fact, a tournament in which any two vertices are dominated by a third. Generalize for n vertices, $n \equiv 3 \pmod 4$.

3. A vertex of a tournament is a "winner" if every other vertex is at distance ≤ 2 from it. Do almost all tournaments have every vertex a winner?

4. Give a "one-line proof" that almost all tournaments are strong. (Hence almost all tournaments are hamiltonian by Camion's theorem [Ca59].)

5. A k-set S of vertices is bad if no other vertex is adjacent to each vertex of S. In Model A with p fixed, how fast can k grow so that we still have no bad k-sets in almost all graphs?

3

THRESHOLD FUNCTIONS

*Most people my age are dead
and you can look it up.*

We return to the interesting question raised earlier that asks for a special function $q = q(n)$ just large enough that in Model B almost all graphs have a particular property. The question is first more carefully formulated and then answered in both Models A and B.

3.1. A THRESHOLD FOR ISOLATED VERTICES

As the number of edges increases in a graph of order n, it is less likely that the graph will have isolated vertices. In fact, if the number of edges is less than $\lfloor n/2 \rfloor$, a graph is certain to have isolated vertices, while if $q > \binom{n-1}{2}$, a graph of order n has none. Is there between $\lfloor n/2 \rfloor$ and $\binom{n-1}{2}$ some special function $t(n)$ such that in Model B if $q = q(n)$ is "bigger than $t(n)$" then almost all graphs have no isolated vertices? Can $t(n)$ satisfy simultaneously the condition that if $q = q(n)$ is "smaller than $t(n)$" then almost every graph is sure to have at least one isolated vertex? Such a function can be defined in a number of ways. The following simple formulation will be adequate for our immediate purpose.

Let Q be a property of graphs, and let $\mathscr{A}$ be the set of graphs of order n and size q with property Q. Then $t = t(c, n)$ is called a *threshold function for property Q* if there is a number c_0 such that in Model B with $q \sim t(c, n)$ and constant c,

(i) $P(\mathscr{A}) \to 1$ if $c > c_0$, that is, almost all graphs have property Q, while, on the other hand,

(ii) $P(\mathscr{A}) \to 0$ if $c < c_0$, that is, almost no graph has property Q.

To illustrate, we let $\mathscr{A}$ be the labeled graphs of order n and size q that have *no* isolated vertices. To find the threshold for this property, we first determine the expected number of isolated vertices in Model B:

$$(3.1.1) \qquad E(X) = \binom{n}{1}\left(\binom{\binom{n-1}{2}}{q}\right)\left(\binom{\binom{n}{2}}{q}\right)^{-1}.$$

A bit of work with Stirling's formula shows that

$$(3.1.2) \qquad \left(\binom{\binom{n-1}{2}}{q}\right)\left(\binom{\binom{n}{2}}{q}\right)^{-1} \sim e^{-2q/n}$$

provided $q = o(n^{3/2})$ [see formula (3.9) of Appendix III]. Hence

$$(3.1.3) \qquad E(X) \sim ne^{-2q/n}.$$

We want to find $q = q(n)$ so that $E(X) \to 0$, that is, almost all graphs have no isolated vertices. With this aim in mind we see that the right side of (3.1.3) is greatly simplified if we just write q as

$$(3.1.4) \qquad q = c\tfrac{1}{2}n \log n,$$

where c is a function of n. But then we have

$$(3.1.5) \qquad E(X) \sim n^{1-c}$$

and an obvious candidate for a threshold function, namely,

$$(3.1.6) \qquad t(c, n) = c\tfrac{1}{2}n \log n$$

where c is a constant. Note that if $q \sim t(c, n)$ in (3.1.6), it does satisfy the condition $q = o(n^{3/2})$ required earlier! Thus

$$(3.1.7) \qquad E(X) \to \begin{cases} 0 & \text{if } c > 1 \\ \infty & \text{if } 0 < c < 1. \end{cases}$$

At this point we know that if $c > 1$, almost all graphs have no isolated vertices. We also know that if $0 < c < 1$ we can expect lots of isolated vertices, but we cannot be certain that almost every graph has at least one isolated vertex. A large number of isolated vertices could be concentrated in a relatively small proportion of graphs.

To show that almost all graphs have isolated vertices when $0 < c < 1$, we use the so-called second-moment method. If $X(G)$ is the number of isolated vertices, we want $P(X = 0) \to 0$. An immediate consequence of Chebyshev's inequality (see Proposition 5.4 of Appendix V) is

$$(3.1.8) \qquad P(X = 0) \le \frac{E(X^2) - E(X)^2}{E(X)^2}.$$

Hence we have only to show that

$$(3.1.9) \qquad E(X^2) \sim E(X)^2$$

to obtain our result.

To find $E(X^2)$, recall the discussion in Section 2.2 and let

$$(3.1.10) \qquad X = X_1 + \cdots + X_n,$$

where $X_i(G) = 1$ or 0 according as the ith vertex of G is isolated or not. On squaring the right side of (3.1.10) and using the linearity of the expectation, we have

$$(3.1.11) \qquad E(X^2) = \sum_{i=1}^{n} E(X_i^2) + \sum E(X_i X_j),$$

where the second sum is over all ordered pairs (i, j) with $i \ne j$.

But since $X_i^2 = X_i$ and all the elements of the second sum are equal, we can write

$$(3.1.12) \qquad E(X^2) = E(X) + n(n - 1)E(X_1 X_2).$$

Now we have only to work out $E(X_1 X_2)$, and we find

$$(3.1.13) \qquad E(X_1 X_2) = \left(\binom{n - 2}{2}_q \right) \left(\binom{n}{2}_q \right)^{-1}.$$

The right side of (3.1.13) is again a form that occurs frequently and has been estimated in formula (3.9) of Appendix III. As a consequence,

$$(3.1.14) \qquad\qquad E(X_1 X_2) \sim e^{-4q/n}.$$

Now the last three formulas can be put together to obtain

$$(3.1.15) \qquad\qquad E(X^2) \sim E(X) + n^2 e^{-4q/n},$$

and on squaring $E(X)$ in (3.1.3) we have

$$(3.1.16) \qquad\qquad \frac{E(X^2)}{E(X)^2} = \frac{1}{E(X)} + 1.$$

Since we are dealing with the case in which $0 < c < 1$, we have $E(X) \to \infty$ [see (3.1.7)], and therefore $E(X^2) \sim E(X)^2$.

These results are summarized in the following theorem.

Theorem 3.1.1. Suppose $q \sim c\frac{1}{2}n \log n$. In Model B with constant $c > 1$, almost all graphs have no isolated vertices, while if $0 < c < 1$, almost all graphs have at least one isolated vertex.

It is important to stress that this result could have been obtained more easily by using Model A. See for yourself in what follows. Let $X(G)$ be the number of isolated vertices in G. Then

$$(3.1.17) \qquad\qquad E(X) = n(1 - p)^{n-1},$$

and we need to find $p = t(c, n)$ and c_0 such that $E(X) \to 0$ or ∞ according as $c > c_0$ or $< c_0$. But this is easy if we just take the log and exponential of $(1 - p)^n$:

$$(3.1.18) \qquad
\begin{aligned}
(1 - p)^n &= \exp\{\log(1 - p)^n\} \\
&= \exp\{n \log(1 - p)\} \\
&= \exp\left\{n\left(-p - \frac{p^2}{2} - \frac{p^3}{3} \cdots\right)\right\} \\
&= \exp\{-np\}\exp\left\{-np^2\left(\frac{1}{2} + \frac{p}{3} + \cdots\right)\right\}.
\end{aligned}$$

Therefore

$$(3.1.19) \qquad\qquad E(X) \sim ne^{-np},$$

provided $np^2 \to 0$.

To simplify the right side of (3.1.19) we should take

(3.1.20)
$$p = c\frac{\log n}{n}.$$

Then, since $np^2 \to 0$,

(3.1.21)
$$E(X) \sim n^{1-c},$$

and $E(X) \to 0$ or ∞ according as $c > 1$ or < 1.

The second-moment method is also easily applied with

(3.1.22)
$$E(X^2) = E(X) + n(n-1)(1-p)^{2(n-2)+1}.$$

As before, $E(X^2) \sim E(X)^2$ for $0 < c < 1$. So equation (3.1.20) provides a threshold in Model A for isolated vertices. Note that we avoided all mucking about with Stirling's formula applied to quotients of binomial coefficients. Furthermore, the threshold given by (3.1.20) for Model A is equivalent to the one derived earlier (3.1.6) for Model B. That is, either of these thresholds can be used to obtain the other by substitution into the formula that equates the expected number of edges in both models:

(3.1.23)
$$q = p\binom{n}{2}.$$

Evidently the two models are very closely related, and so in many cases we will choose to work in Model A to simplify the computations (see Appendix VI for further details on the relationship between the models).

We demonstrate the method one more time by finding a threshold that guarantees that almost every graph as a given graph H as a subgraph. For H we take the unique graph of order 4 and size 5, i.e., H is the next-to-last graph in Figure 1.1.1. Let $X(G)$ be the number of subgraphs of G that are isomorphic to H. Then the four-element subsets of the n vertices are listed from 1 to $t = \binom{n}{4}$, and X is written as the sum

(3.1.24)
$$X = X_1 + \cdots + X_t,$$

where $X_i(G)$ is the number of subgraphs with vertices in the ith subset that are isomorphic to H. Then, just as in Section 2.2, we have for Model A

(3.1.25)
$$\begin{aligned}
E(X) &= \sum_{i=1}^{t} E(X_i) \\
&= \binom{n}{4}E(X_1) \\
&= \binom{n}{4}\frac{4!}{4}p^5,
\end{aligned}$$

where the factor $4!/4$ is the number of ways to label H. Therefore the asymptotic behavior of $E(X)$ is

$$(3.1.26) \qquad E(X) = \frac{(n)_4 p^5}{4} \sim \frac{n^4 p^5}{4},$$

and so $E(X) \to 0$ or ∞ according as $np^{5/4} \to 0$ or ∞. The corresponding result in Model B is

$$(3.1.27) \qquad E(X) = \binom{n}{4} \frac{4!}{4} \left(\frac{\binom{n}{2} - 5}{q - 5} \right) \left(\frac{\binom{n}{2}}{q} \right)^{-1},$$

and from formula (3.11) of Appendix III we can show that

$$(3.1.28) \qquad E(X) \sim \frac{1}{2} \frac{q^5}{n^6}.$$

In this case $E(X) \to 0$ if $q = o(n^{6/5})$, and $E(X) \to \infty$ when $q = \omega_n n^{6/5}$ for some $\omega_n \to \infty$ arbitrarily slowly.

Now we know that in Model A almost no graph has H as a subgraph if $np^{5/4} \to 0$, but we must use the second-moment method to show that if $np^{5/4} \to \infty$, almost every graph has a subgraph isomorphic to H. Hence we need

$$(3.1.29) \qquad E(X^2) = E(X) + \sum E(X_i X_j)$$

where the sum is over all ordered pairs of four element subsets of the n vertices. There are four types of contributions to this sum, which depend on the number m of vertices common to both pairs of subsets. The dominant contribution is made when $m = 0$. It is

$$(3.1.30) \qquad \binom{n}{4, 4, n - 8} \left(\frac{4!}{4} \right)^2 p^{10} \sim \left(\frac{n^4 p^5}{4} \right)^2.$$

The multinomial coefficient in (3.1.30) selects the ordered pair of subsets, $(4!/4)^2$ is the number of ways they can be filled with copies of H and p^{10} assures that the 10 edges of the two copies are present.

After considering the numerous ways in which two copies of H intersect in 1, 2 or 3 vertices, we find the contributions for $m = 1, 2$ and 3 are

$$\binom{n}{3, 1, 3, n - 7} 36 p^{10} + \binom{n}{2, 2, 2, n - 6} (11 p^{10} + 25 p^9)$$
$$+ \binom{n}{1, 3, 1, n - 5} (6 p^9 + 21 p^8 + 9 p^7).$$

The exact formula for $E(X^2)$ is

$$(3.1.31) \quad \begin{aligned} E(X^2) &= E(X) + (n)_8 \frac{p^{10}}{16} + (n)_7 p^{10} + \frac{(n)_6(11p^{10} + 25p^9)}{8} \\ &+ \frac{(n)_5(2p^9 + 7p^8 + 3p^7)}{2}. \end{aligned}$$

If $n^4 p^5 \to \infty$, then it is easy to check the terms in (3.3.31) to see that $E(X^2)/E(X)^2 \sim 1$. Hence almost every graph has a subgraph isomorphic to H.

A graph is *balanced* if its average degree is at least as large as the average degree of any of its subgraphs. Erdös and Rényi [ErR60] established a threshold for the appearance of any balanced graph as a subgraph of almost all graphs.

Theorem 3.1.2. Let H be any connected balanced graph of order $k \geq 2$ and size l with $k - 1 \leq l \leq \binom{k}{2}$, and let $\mathscr{A}$ be the set of graphs of order n that contain H as a subgraph. Then in Model A,

$$(3.1.32) \quad P(\mathscr{A}) \to \begin{cases} 0 & pn^{k/l} \to 0 \\ 1 & pn^{k/l} \to \infty, \end{cases}$$

and in Model B, where $\mathscr{A}$ consists of graphs of order n and size q that contain H as a subgraph, the result is

$$(3.1.33) \quad P(\mathscr{A}) \to \begin{cases} 0 & q = \dfrac{n^{2-k/l}}{\omega_n} \\ 1 & q = \omega_n n^{2-k/l}, \end{cases}$$

where $\omega_n \to \infty$ arbitrarily slowly.

The proof is carried out in exactly the same way we investigated the graph of order 5 and size 4. The requirement of a balanced graph in the hypothesis ensures that $E(X^2) \sim E(X)^2$. For example, consider the unbalanced graph H of order 5 and size 7 obtained by adding an edge to a complete graph of order 4. This graph has average degree $14/5$ and a subgraph K_4 with higher average degree 3. As before, $E(X)$ is the expected number of subgraphs isomorphic to H, and so $E(X) \to \infty$ if $\omega_n = n^5 p^7 \to \infty$. But consider the contribution to $E(X^2)$ made by two copies of H that intersect exactly at the complete subgraphs of order 4. This contribution will be approximately $n^6 p^8$. To ensure that $E(X^2) \sim E(X)^2$, we will need

(3.3.34)
$$\frac{n^6 p^8}{(n^5 p^7)^2} \to 0.$$

But if we solve $\omega_n = n^5 p^7$ for p and substitute the result in (3.1.34) we find

(3.1.35)
$$\frac{n^6 (\omega_n/n^5)^{8/7}}{\omega_n^2} = (n^2 \omega_n^{-6})^{1/7} \to \infty,$$

a contradiction.

Here are the crucial steps of the proof of Theorem 3.1.2. The random variable $X(G)$ counts the number of subgraphs of G that are isomorphic to the balanced graph H. The expectation is

(3.1.36)
$$E(X) = \binom{n}{k} \frac{k!}{s(H)} p^l \sim \frac{n^k p^l}{s(H)},$$

where $k!/s(H)$ is the number of ways to label H. For the second moment we have

(3.1.37)
$$E(X^2) = E(X) + \binom{n}{k, k, n - 2k}\left(\frac{k!}{s(H)}\right)^2 p^{2l} + \cdots$$
$$+ \binom{n}{k - m, m, k - m, n - 2k + m} M_m p^{2l-r} + \cdots.$$

The typical term in (3.1.37) is contributed by intersections of two copies of H in which the intersection is a subgraph H_1 of order m and size r. The factor M_m is just the number of labelings of the vertices of the ordered pair of copies. This typical term is asymptotic to

(3.1.38)
$$\frac{M_m}{[(k - m)!]^2 m!} n^{2k-m} p^{2l-r}.$$

To prove that $E(X^2) \sim E(X)^2$ when $E(X) \to \infty$, we divide (3.1.38) by the square of the right side of (3.1.36) to obtain a constant times $n^{-m} p^{-r}$. Now we need this expression $n^{-m} p^{-r}$ to go to zero as $n \to \infty$ or we need $pn^{m/r} \to \infty$. But $2r/m$ is the average degree of the subgraph H_1 of H. Since H is balanced,

(3.1.39)
$$\frac{2r}{m} \le \frac{2l}{k},$$

and therefore

$$(3.1.40) \qquad\qquad pn^{k/l} \le pn^{m/r}.$$

But by hypothesis the left side of (3.1.40) goes to $+\infty$, and hence so does $pn^{m/r}$.

The theorem can be applied to trees, cycles, connected unicyclic graphs and complete graphs.

Exercises 3.1

1. In Model A with $p = c(\log n)/n$ and $c < 1$, use Chebyshev's inequality to show that almost every graph has at least m isolated vertices where $m = o(n^{1-c})$.

2. Use Chebyshev's inequality to show that in Model A with $pn^2 \to \infty$ and any $\varepsilon > 0$, the number $X(G)$ of edges of almost every graph satisfies

$$(1 - \varepsilon)p\binom{n}{2} < X(G) < (1 + \varepsilon)p\binom{n}{2}.$$

3. Find a threshold for isolated edges in Model A and in Model B.

4. Let $X(G)$ be the number of edges of G that are not in triangles. Find $E(X^2)$ in Model A.

5. Find a threshold in Model A for the property that every edge is in a triangle, and provide a complete proof.

6. In exercise 2.3.5, how fast should k grow so that we are sure to have a bad k-set in almost every graph?

3.2. A SHARPER THRESHOLD

The threshold function found in (3.1.20) for isolated vertices suggests another question. What is the likelihood that a random graph has isolated vertices if $c = 1$? To answer this we replace c by $1 + \varepsilon_n$, where $\varepsilon_n = o(1)$ so that

$$(3.2.1) \qquad\qquad p = (1 + \varepsilon_n)\frac{\log n}{n}.$$

Then the expected number of isolated vertices will be

$$(3.2.2) \qquad\qquad E(X) \sim n^{-\varepsilon_n},$$

and we can define ε_n in any convenient way we choose. One sensible approach is to introduce a new variable x and select ε_n so that $E(X) \sim e^{-x}$. To do this we just set $e^{-x} = n^{-\varepsilon_n}$ and solve for ε_n:

$$(3.2.3) \qquad\qquad \varepsilon_n = \frac{x}{\log n}.$$

Summarizing, we have $E(X) \sim e^{-x}$ when $p = (\log n)/n + x/n$. As x ranges from $-\infty$ to $+\infty$, the probability of an edge increases and the expected number of isolated vertices ranges from $+\infty$ to 0. Now we want to know the probability that a graph has no isolated vertices, that is, $P(X = 0)$. This probability can be determined by the method of inclusion and exclusion, and, in fact, for any k we can find $P(X = k)$ from Jordan's formula (see Proposition 5.8 of Appendix V). As in formula (3.1.10), $X_i(G) = 1$ or 0 according as the ith vertex of G is isolated or not. It remains to evaluate for each r

$$(3.2.4) \qquad\qquad S_r = \sum E(X_{l_1} \cdots X_{l_r})$$

where the sum is over all sequences $1 \le l_1 < \cdots < l_r \le n$.

Of course, by definition $S_0 = 1$, and we already know from formulas (3.1.17) and (3.1.22) that

$$(3.2.5) \qquad\qquad S_1 = E(X) = \binom{n}{1}(1 - p)^{n-1}$$

and

$$(3.2.6) \qquad\qquad \begin{aligned} S_2 &= \frac{E(X^2) - E(X)}{2} \\ &= \binom{n}{2}(1 - p)^{2(n-2)+1}. \end{aligned}$$

It is easy to see that for any r,

$$(3.2.7) \qquad\qquad S_r = \binom{n}{r}(1 - p)^{r(n-r)+\binom{r}{2}},$$

and so for fixed r, since $p \to 0$,

$$(3.2.8) \qquad\qquad S_r \sim \frac{\left[n(1 - p)^n\right]^r}{r!}.$$

But our special choice of p in (3.2.1) and (3.2.3) implies

$$(3.2.9) \qquad\qquad S_r \sim \frac{(e^{-x})^r}{r!}.$$

Finally the Bonferroni inequalities can be applied to estimate $P(X = k)$ [see formula (5.22) of Appendix V], and we summarize these results in the following statement.

Theorem 3.2.1. In Model A with the probability of an edge given by

$$(3.2.10) \qquad p = \frac{\log n}{n} + \frac{x}{n},$$

the random variable X that counts isolated vertices is distributed according to Poisson's law; that is, for each $k = 0, 1, 2, \ldots$,

$$P(X = k) \to \frac{e^{-\mu}\mu^k}{k!}$$

with mean $\mu = e^{-x}$.

A similar conclusion holds for Model B (see exercise 3.2.2).

Here is a slightly more complicated illustration. Let $X(G)$ be the number of components of G that are isomorphic to the little bipartite graph $K_{1,3}$. First let us find a threshold for the disappearance of these graphs as components. The expectation of X in Model A is

$$(3.2.11) \qquad \begin{aligned} E(X) &= \binom{n}{4}4p^3(1-p)^{3+4(n-4)} \\ &\sim \frac{n^4}{3!}p^3(1-p)^{4n}, \end{aligned}$$

provided $p \to 0$. To estimate $(1-p)^{4n}$, we use the same old formulas (3.1.18) and (3.1.20). With $p = c(\log n)/n$, we have $(1-p)^n \sim n^{-c}$, and therefore

$$(3.1.12) \qquad E(X) \sim \frac{c^3}{3!}\frac{(\log n)^3}{n^{4c-1}}.$$

Then $E(X) \to 0$ for $c > \frac{1}{4}$, and almost no graph has $K_{1,3}$ as a component. For $c \le \frac{1}{4}$, $E(X) \to \infty$, and the second-moment method can be applied to show that almost every graph has a component isomorphic to $K_{1,3}$.

How does $E(X)$ behave if $c \to \frac{1}{4}$ as $n \to \infty$? We substitute $c = \frac{1}{4} + \varepsilon_n$ for c in the right side of (3.2.12), set the result equal to e^{-x} and take logs:

$$(3.2.13) \qquad -x = 3\log(1 + 4\varepsilon_n) - \log 384 + 3\log\log n - 4\varepsilon_n \log n.$$

Now we replace $3\log(1 + 4\varepsilon_n)$ by $o(1)$ and solve for ε_n. The result is

$$(3.2.14) \qquad \varepsilon_n = \frac{3\log\log n - \log 384 + x + o(1)}{4\log n}.$$

Note that $\varepsilon_n \to 0$ as required.

With ε_n defined by (3.2.14) and

$$(3.2.15) \qquad p = \left(\tfrac{1}{4} + \varepsilon_n\right)\frac{\log n}{n},$$

we need to check that

$$(3.2.16) \qquad E(X) \sim e^{-x}$$

for any real number x. This can be done in just a few steps. To finish the job we need only work out the binomial moments. By definition, $S_0 = 1$, and we have just shown that $S_1 = E(X) \sim e^{-x}$. Using definition (5.13) of Appendix V for $r \geq 1$, we find

$$(3.2.17) \quad S_r = \frac{1}{r!}\binom{n}{4,\ldots,4,\,n-4r}4^r(p^3)^r(1-p)^{\binom{4r}{2}-3r+4r(n-4r)}$$

and so

$$(3.2.18) \qquad \begin{aligned} S_r &\sim \frac{1}{r!}\left[\frac{n^4}{3!}p^3(1-p)^{4n}\right]^r \\ &\sim \frac{(e^{-x})^r}{r!}. \end{aligned}$$

Once again the Bonferroni inequalities (see Appendix V) show that this random variable obeys Poisson's law. Obviously the outcome will be similar for the distribution of any specified component.

Exercises 3.2

1. Plot the curve $y = \exp[-\exp(-x)]$ of the limiting probability in Theorem 3.2.1 when $k = 0$.

2. Let X count isolated vertices. Find a function $t(x, n)$ so that if $q \sim t(x, n)$ then $E(X) \sim e^{-x}$ in Model B.

3. Find a sharp threshold for isolated edges in Model A.

4. Let X count edges that are not in triangles. Find $p = t(n, x)$ so that $E(X) \sim e^{-x}$ in Model A.

5. Find a sharp threshold for the property that a graph has diameter ≤ 2. That is, let $\mathcal{D}$ be the set of graphs G of order n such that the distance between any two vertices of G is at most 2. Find $p = t(n, x)$ so that in Model A

$$P(\mathcal{D}) \to \exp[-\exp(-x)]$$

for any real number x.

3.3. THRESHOLDS FOR EXISTENCE

Often the simplest concepts from probability theory can be used to prove the existence of certain configurations that are quite difficult to construct. Suppose we want to show that a graph with some particular property Q exists. Let $\mathscr{A}$ be the set of all graphs of order n that have this property. If we can establish that $P(\mathscr{A}) > 0$, then $\mathscr{A} \neq \varnothing$, and so such a graph must exist. Often our graphical events are described by a nonnegative, integer-valued random variable $X = X(G)$. To prove that there is a graph G with $X(G) < t$ for some t, we try to show that

$$(3.3.1) \qquad \frac{E(X)}{t} < 1.$$

Then Markov's inequality (Proposition 5.2 of Appendix V) can be applied to obtain

$$(3.3.2) \qquad P(X \geq t) \leq \frac{E(X)}{t} < 1,$$

and therefore

$$(3.3.3) \qquad P(X < t) > 0.$$

Hence there exists a graph G with $X(G) < t$. If $E(X) < 1$, then there is a graph G with $X(G) = 0$. Shütte's problem was solved in Section 2.3 by showing even more, namely, that $E(X) \to 0$. Hence it was found that *almost all* tournaments T had the property $X(T) = 0$.

In general, for Models A, B and C, if there is a function $t = t(n)$ such that

$$(3.3.4) \qquad E(X) = o(t(n)),$$

then

$$(3.3.5) \qquad P(X \geq t(n)) \leq \frac{E(X)}{t(n)} = o(1),$$

and so almost every graph G has $X(G) < t(n)$.

The earliest significant use of this nonconstructive method is found in the work of Erdös [Er47], where lower bounds for Ramsey numbers are established. The Ramsey number $r(k, l)$ is the smallest integer such that every graph of order $n \geq r(k, l)$ has either the complete graph K_k or the complement $\overline{K}_l$ as a subgraph. A few of the small Ramsey numbers are known. For example, $r(3, 3) = 6$ and $r(4, 4) = 18$, but $r(5, 5)$ has not yet been discovered.

To determine a lower bound, say $n \leq r(k, l)$, we just need to prove that a graph of order n exists that has neither K_k nor $\overline{K}_l$ as subgraphs. Let $X(G)$ be the number of subgraphs of G that are isomorphic to K_k or $\overline{K}_l$. Then in Model A, as noted by Spencer [Sp78],

$$(3.3.6) \qquad E(X) = \binom{n}{k} p^{\binom{k}{2}} + \binom{n}{l}(1-p)^{\binom{l}{2}},$$

and for $k = l$ and $p = 1/2$

$$(3.3.7) \qquad E(X) = \binom{n}{k} 2^{1 - \binom{k}{2}}.$$

The theorem of Erdös follows instantly from our earlier observation that $E(X) < 1$ implies $P(X = 0) > 0$.

Theorem 3.3.1. If $\binom{n}{k} 2^{1-\binom{k}{2}} < 1$, then $r(k, k) > n$.

An easy application of Stirling's formula produces a lower bound in terms of k.

Corollary 3.3.1

$$r(k, k) > \frac{k 2^{k/2}}{e\sqrt{2}}.$$

Spencer [Sp78] used the Lovász local lemma [ErL75] to improve this bound by a factor of 2.

Volume 7 (1983) of the *Journal of Graph Theory* was dedicated to the memory of Frank P. Ramsey, and it is in that issue that the reader can find the latest results on Ramsey numbers and Ramsey theory. In particular, the article [ChG83] by Chung and Grinstead can be consulted for tabulated values as well as bounds for graphical Ramsey numbers. For generalized Ramsey theory, there is the book [GrRS80] by Graham, Rothschild and Spencer.

In this section our main objective is to deal with a later result [Er59] that is justly famous for its power and glory. The treatment here follows that of [ErS74] and [Sp78].

If a graph G has a complete graph K_m as a subgraph, then the chromatic number of G is at least m, that is, $\chi(G) \geq m$. This raises the question of whether or not there is a graph with chromatic number $m \geq 3$ but no triangles as subgraphs. Erdös answered in the affirmative with a much stronger result.

The *girth* of a graph is the order of a smallest cycle. The disjoint union of two cycles of order $g \geq 3$ and $g + 1$ has girth g and $\chi = 3$. The pioneering existence theorem of Erdös [Er59] settles the nontrivial cases for $\chi \geq 4$.

Here is a rough guide to the proof. Our goal is to find a graph with high chromatic number and high girth. If we locate a graph with high χ and not too many small cycles, then one edge from each cycle could be deleted to increase the girth. But when an edge e is deleted from a graph G, the chromatic number remains the same or drops by 1:

$$(3.3.8) \qquad \chi(G) \geq \chi(G - e) \geq \chi(G) - 1.$$

Hence preemptive steps must be taken to ensure that χ is not reduced when the girth is increased by edge deletion. To keep the chromatic number high despite the deletion of edges, we use the inequality

$$(3.3.9) \qquad \chi \geq \frac{n}{\beta},$$

where β is the independence number, that is, the maximum number of mutually nonadjacent vertices. If every r-subset of vertices of G has at least $t = t(n)$ edges, we can delete $t(n) - 1$ edges and still have $\beta < r$ and hence $\chi > n/r$.

Theorem 3.3.2. For every fixed pair of integers $m \geq 4$ and $g \geq 4$, there is a graph with chromatic number m and girth g.

Proof. We work in Model A with two random variables, one to keep track of small cycles and the other to measure the lower bound n/β for the chromatic number. Let $X(G)$ be the number of cycles in G of order less than g. Then

$$
\begin{aligned}
E(X) &= \sum_{k=3}^{g-1} \binom{n}{k} \frac{(k-1)!}{2} p^k \\
(3.3.10) \qquad &\leq \sum_{k=3}^{g-1} (pn)^k \\
&\leq (g-1)(pn)^{g-1}
\end{aligned}
$$

for any $pn \geq 1$. For any function of n, say $t = t(n)$, such that

$$(3.3.11) \qquad (pn)^{g-1} = o(t),$$

it follows from Markov's inequality that

$$(3.3.12) \qquad\qquad P(X \geq t) \to 0,$$

that is, almost every graph has fewer than t cycles of order less than g.

The second random variable $Y(G)$ counts the number of r-subsets of vertices of G with fewer than t edges, and $r = r(n)$ is another function to be determined. The expectation of Y is seen to be

$$(3.3.13) \qquad E(Y) = \binom{n}{r} \sum_{k=0}^{t-1} \binom{\binom{r}{2}}{k} p^k (1-p)^{\binom{r}{2}-k}.$$

We need to find p, r and t so that (3.3.11) holds, $E(Y) \to 0$ and $n/r \to \infty$. Then we can say that almost every graph has

(i) fewer than t cycles of order less than g,

(ii) at least t edges in every r-subset of vertices,

(iii) $\chi \geq n/r$.

Let G be such a graph. The proof could then almost be finished by deleting an edge from each cycle of order less than g.

To determine p, r and t we require a good estimate of $E(Y)$:

$$(3.3.14) \qquad E(Y) \leq \binom{n}{r} \binom{\binom{r}{2}}{t-1} (1-p)^{\binom{r}{2}} \sum_{k=0}^{t-1} \left(\frac{p}{1-p}\right)^k$$

provided $t - 1 \leq \binom{r}{2}/2$.

If $p \to 0$ we have

$$(3.3.15) \qquad E(Y) \leq \binom{n}{r} \binom{\binom{r}{2}}{t} (1-p)^{\binom{r}{2}} [1 + o(1)],$$

and from formula (3.1) of Appendix III it follows that

$$(3.3.16) \qquad E(Y) \leq n^r \left(\frac{er^2}{2t}\right)^t (1-p)^{\binom{r}{2}}.$$

This last inequality can be simplified a bit more by using the fact that

$$(3.3.17) \qquad\qquad (1-p) < e^{-p},$$

to obtain

$$(3.3.18) \qquad E(Y) \leq n^r \left(\frac{er^2}{2t} \right)^t e^{-p\binom{r}{2}}.$$

It is a bit easier to solve for p, r and t if we assume

$$(3.3.19) \qquad p \sim n^{-\alpha}, \qquad r \sim n^\theta, \qquad t \sim n^\gamma$$

for constants $0 < \alpha < 1$, $0 < \theta < 1$ and $0 < \gamma < 2$. If we substitute these on the right side of (3.3.18) and take logarithms, we find

$$(3.3.20) \quad \log E(Y) \leq n^\theta \log n + n^\gamma \log en^{2\theta} + \frac{n^{\theta-\alpha}}{2} - n^\gamma \log 2n^\gamma - \frac{n^{2\theta-\alpha}}{2}.$$

Now it is easy to see that $\log E(Y) \rightarrow -\infty$ and also (3.3.11) holds if

$$(3.3.21) \qquad 2\theta - \alpha > \max\{\gamma, \theta\}$$

and

$$(3.3.22) \qquad \gamma > (1-\alpha)(g-1).$$

There are many solutions for these constraints, but for simplicity we take $\gamma = 1$ and hence $t = n$. By definition, $\theta < 1 = \gamma$, and so (3.3.21) is satisfied if $2\theta - \alpha > 1$. Finally, just choose α and θ so that

$$(3.3.23) \qquad 1 - \frac{1}{g-1} < \alpha < 2\theta - 1 < 1.$$

Thus p, r and t can be defined by

$$(3.3.24) \qquad p = n^{-\alpha}, \qquad r = \lfloor n^\theta \rfloor, \qquad t = n.$$

At this point we know that in Model A with p, r and t defined by (3.3.24), almost all graphs satisfy conditions (i) and (ii) and have $\chi \geq n^{1-\theta}$. Therefore, for n sufficiently large, there is a graph G_1 that satisfies (i) and (ii) and has $\chi(G_1) \geq n^{1-\theta} \geq m$. Delete from G_1 an edge from each cycle of order less than g to obtain a graph G_2 with girth at least g and chromatic number at least m. Delete more edges from G_2 until a new graph G_3 is obtained whose chromatic number has been reduced to m. Finally, add a component to G_3 that is a cycle of order g to get a graph G_4 with chromatic number m and girth g. $\square$

The theorem is of no help in constructing examples (see exercise 3.3.3). It was almost ten years before Lovász showed how this could be done [Lo68].

More material and references on this subject including applications to coding theory can be found in the expository article on nonconstructive methods by Joel Spencer [Sp78].

Exercises 3.3

1. Let X be a random variable defined for graphs of order n. Show that there is a graph G with $X(G) \geq E(X)$, that is, not every graph is below average.

2. Show that for every $n \geq 3$ there is a labeled tournament of order n with at least $(n - 1)!2^{-n}$ hamiltonian cycles. Find a tournament of order 5 with at least three hamiltonian cycles.

3. Construct a graph with chromatic number and girth 4.

4. Supply the details of the proof of Corollary 3.3.1.

5. Use the probabilistic method to find a lower bound for the Ramsey number $r(5, 6)$.

4

THE EVOLUTION OF RANDOM GRAPHS

3 Billion B.C.

The Earth is a swirling ball of flaming gases.
Fishing is extremely poor, especially in August.

CLIFF HAUPTMAN
The Complete History of Fishing

As suggested by Erdös and Rényi [ErR60], a random graph of large order n can be thought of as a configuration that begins its development with no edges at all and evolves by acquiring more and more edges in some random fashion. Their amazing discovery was that there are certain structural features of the evolving graph that occur abruptly.* For example, we have seen a rather precise description of the stage at which the isolated vertices vanish. Erdös and Rényi used Model B to determine many other thresholds for significant structural properties of almost all graphs. Some of these results are quite unexpected and even surprising. In this chapter we will discuss three of the most important phases of growth.

The fact that random graphs undergo sudden structural changes as the number of edges increases has not gone unnoticed by chemists who study phase transition in chemical systems. There are current research efforts [ErK-U] to refine the Erdös-Rényi theory to serve as a model for predicting the abrupt

---

*The paleontologists call this phenomenon Punctuated Equilibrium or Punk Eek, for short.

changes from vapor to liquid to solid for polymers subjected to temperature reduction.

4.1. THE EMERGING FORESTS

When the evolving random graph first acquires edges, these edges will be widely scattered in small trees. For this range of development, the probability of an edge in Model A should be so small that almost all graphs have no cycles. Therefore we begin our analysis by considering $X(G)$, the number of cycles in G, and we find for Model A

$$(4.1.1) \qquad E(X) = \sum_{k=3}^{n} \binom{n}{k} \frac{(k-1)!}{2} p^k.$$

Immediately we have the upper bound:

$$(4.1.2) \qquad E(x) \leq \sum_{k=3}^{n} \frac{(pn)^k}{2k},$$

from which it follows that if $pn \to 0$, so does $E(X)$, and we are assured that almost no graph has cycles. In Model B, the corresponding requirement that $q = o(n)$ is not quite so instantly obtained (see exercise 4.1.1).

Erdös and Rényi found out much more about the behavior of $E(X)$ (see [ErR60]). For example, equation (4.1.1) can be rewritten:

$$(4.1.3) \qquad E(X) = \frac{1}{2} \sum_{k=3}^{n} \frac{(n)_k}{n^k} \frac{(pn)^k}{k}.$$

Then clearly the next level one ought to consider is constant pn. We use $pn = 2c$, where $c > 0$ is constant to remain consistent with [ErR60]. By inspection, $E(X)$ will converge if $2c < 1$ and will diverge if $2c = 1$. The exact behavior of $E(X)$ hinges on the behavior of $(n)_k/n^k$, which is well documented in Appendix III. Here is a sketch of the analysis.

(Case i) $0 < c < 1/2$: Since $(n)_k/n^k = 1 + o(1)$ for $k = o(n^{1/2})$, we have

$$(4.1.4) \qquad E(X) = \frac{1}{2} \sum_{k=3}^{\sqrt{n}/\omega_n} [1 + o(1)] \frac{(2c)^k}{k} + O(1) \sum_{k > \sqrt{n}/\omega_n}^{n} \frac{(2c)^k}{k}$$

where $\omega_n \to \infty$ slowly. The big-O term is the remainder of a convergent series (since $2c < 1$), and hence we have

$$(4.1.5) \qquad E(X) \sim \frac{1}{2} \sum_{k=3}^{\infty} \frac{(2c)^k}{k} = -\tfrac{1}{2} \log(1 - 2c) - c - c^2.$$

(Case ii) $c = 1/2$: Equation (4.1.4) still holds, and the first sum on the right side is estimated by

$$(4.1.6) \qquad \frac{1}{2} \sum_{k=1}^{\sqrt{n}/\omega_n} \frac{1}{k} \sim \frac{1}{2} \log\left(\frac{\sqrt{n}}{\omega_n}\right)$$

which in turn is asymptotic to $\frac{1}{4}\log n$ provided $\log \omega_n = o(\log n)$. But the big-$O$ term in (4.1.4) must be more closely inspected. We use a better estimate of $(n)_k/n^k$ and write

$$(4.1.7) \qquad \sum_{k>\sqrt{n}/\omega_n}^{n} \frac{(n)_k}{n^k}\frac{1}{k} = O\left(\sum_{k>\sqrt{n}/\omega_n}^{\sqrt{n}\,\omega_n} \frac{1}{k} + \sum_{k>\sqrt{n}\,\omega_n}^{n} \frac{1}{k}e^{-k^2/2n}\right)$$

$$= O\left(2 \log \omega_n + e^{-\omega_n^2/2}\log n\right).$$

Now if we choose $\omega_n = \log n$, the right side of (4.1.7) is $o(\log n)$ and

$$(4.1.8) \qquad E(X) \sim \tfrac{1}{4}\log n.$$

The approach taken in Section 3.2 can also be used to obtain the following result of [ErR60].

Theorem 4.1.1. In Model A with $p = 2c/n$, $c \le 1/2$ and $X(G)$ the number of cycles of G, the limiting probability of a cycle is

$$(4.1.9) \qquad P(X \ne 0) \to 1 - \sqrt{1 - 2c}\,e^{c+c^2}.$$

Note that the limiting value on the right side of (4.1.9) is, as usual, $1 - e^{-\lambda}$, where $\lambda \sim E(X)$ for $c < 1/2$, and the value is 1 when $c = 1/2$. So we are not guaranteed a cycle until $c = 1/2$ or $p = 1/n$. Thus the emerging forests are confined to the stage of development where $pn \to 0$.

Now we want to know what the probability of an edge should be to guarantee that a tree of order k should appear as a component. Let $X(G)$ be the number of components of G that are trees of order k. Then

$$(4.1.10) \qquad E(X) = \binom{n}{k} k^{k-2} p^{k-1}(1-p)^{\binom{k}{2}-(k-1)+k(n-k)}.$$

Note that Cayley's formula [C89] for the number k^{k-2} of labeled trees plays an important role here. Since k is fixed and $pn \to 0$, there is some simplification asymptotically:

$$(4.1.11) \qquad E(X) \sim \frac{n^k}{k!}k^{k-2}p^{k-1}.$$

Thus if we define $p = 2cn^{-k/(k-1)}$, then we can say that $E(X) \to \mu$ with $\mu = (2c)^{k-1}k^{k-2}/k!$.

Now the same approach as was used in Chapter 3 for isolated vertices can be used to show that the distribution of X in the limit follows Poisson's law. As before, the essential steps require the proof that $S_r \sim \mu^r/r!$ for each r. The final result is summarized in the following theorem.

Theorem 4.1.2. In Model A with $p = 2cn^{-k/(k-1)}$, the random variable X, which counts trees of order k that are components, is distributed, in the limit, by Poisson's law with mean μ. Specifically,

$$(4.1.12) \qquad P(X = j) \to e^{-\mu}\frac{\mu^j}{j!}$$

where

$$(4.1.13) \qquad \mu = (2c)^k \frac{k^{k-2}}{k!}.$$

If $pn^{k/(k-1)} \to \infty$ so slowly that we still have $pn = o(1)$, then the second-moment method can be used to show that any tree of order k is sure to be a component of almost every graph (see exercise 4.1.3). In fact, the distribution of the components of order k approaches a normal distribution, and we refer the reader to [ErR60] for the details.

SUMMARY

Model A	Model B	Description of Almost All Graphs
$pn = o(1)$	$q = o(n)$	There are no cycles. Hence, all components are trees.
$pn^{k/(k-1)} = o(1)$	$q = o(n^{(k-2)/(k-1)})$	There are no components of order at least k.
$pn^{k/(k-1)} = 2c$	$q \sim cn^{(k-2)/(k-1)}$	The trees of order k are distributed according to Poisson's law with mean $\mu = (2c)^k k^{k-2}/k!$
$\omega_n \to \infty$ arbitrarily slowly $pn^{k/(k-1)} = \omega_n$	$q = \omega_n n^{(k-2)/(k-1)}$	Every tree of order k is a component.

Exercises 4.1

1. Find the expected number of cycles using Model B.

2. Use Model A with $pn \to \infty$ to show that almost every graph has a triangle.

3. Prove that if $pn^{4/3} \to \infty$ slowly, almost every graph has a path of order 4 as a component.

4. If $pn^{k/(k-1)} = o(1)$ for $k \geq 2$, show that almost no graph has components of order $\geq k$.

4.2. THE DOUBLE JUMP

The second phase of growth that we will study occurs when the probability of an edge is high enough that cycles may appear. We have seen that in Model A with $p = 2c/n$ (or equivalently in Model B with $q \sim cn$), the limiting probability of a cycle of order k is positive and in fact is 1 when $c = 1/2$. This value of c is also critical for estimating the number of vertices of a largest component in a random graph. One of the most surprising results found by Erdös and Rényi was that the order of a largest component jumps dramatically as c passes $1/2$. For $c < 1/2$ this order is about $\log n$, but it jumps to about $n^{2/3}$ for $c = 1/2$. For $c > 1/2$, the order jumps again to about n in magnitude. These comments are explicitly formulated in the theorems of this section.

We begin the investigation in Model A by defining $X_m(G)$ to be the number of components of G that are trees of order at least m. Then the expected value is

$$(4.2.1) \qquad E(X_m) = \sum_{k=m}^{n} \binom{n}{k} k^{k-2} p^{k-1} (1-p)^{\binom{k}{2} - (k-1) + k(n-k)}.$$

Our task now, with $p = 2c/n$ and constant $c > 0$, is to find m, as a function of n and c, so low that we still have $E(X_m) \to 0$ as $n \to \infty$, that is, almost no graph has a component that is a tree of order $\geq m$.

We use the following approximations:

$$(4.2.2) \qquad (1-p)^{\binom{k}{2} - (k-1) + k(n-k)} = O(1) \left(\frac{e^{kc/n}}{e^{2c}} \right)^k$$

$$(4.2.3) \qquad \binom{n}{k} = O(1)(n)_k \frac{e^k}{k^{k+1/2}},$$

and on substitution in (4.2.1) we find

(4.2.4) $\qquad E(X_m) = O(1)n \sum_{k=m}^{n} \frac{(n)_k}{n^k} \frac{1}{k^{5/2}} \left[(e^c)^{k/n} (2ce^{1-2c}) \right]^k.$

For $0 < c \le 1/2$ we use the bound from Appendix III:

(4.2.5) $\qquad\qquad\qquad \frac{(n)_k}{n^k} = O(1)(e^{-1/2})^{k^2/n}$

together with the fact that

(4.2.6) $\qquad\qquad\qquad (c - 1/2)\frac{k^2}{n} \le 0.$

Then we have

(4.2.7) $\qquad\qquad\qquad \frac{(n)_k}{n^k} e^{ck^2/n} = O(1),$

and on substitution in (4.2.4) we obtain

(4.2.8) $\qquad\qquad E(X_m) = O(1)n \sum_{k=m}^{n} \frac{1}{k^{5/2}} (2ce^{1-2c})^k.$

For $c = 1/2$ we have

(4.2.9) $\qquad\qquad\qquad E(X_m) = O(1)n \sum_{k=m}^{n} \frac{1}{k^{5/2}},$

and on approximating this sum by an integral, we find that $E(X_m) \to 0$ for $m = \omega_n n^{2/3}$, where $\omega_n \to +\infty$ arbitrarily slowly.

It can then be shown that $E(X_m^2) \sim E(X_m)^2$ for $m = n^{2/3}/\omega_n$ and $c = 1/2$. These results are summarized below.

Theorem 4.2.1. In Model A with $p = 1/n$ and X_m the random variable that counts trees of order at least m that are components,

(4.2.10) $\qquad\qquad P(X_m \ge 1) \to 0 \qquad$ for $m = \omega_n n^{2/3}$,

but

(4.2.11) $\qquad\qquad P(X_m \ge 1) \to 1 \qquad$ for $m = n^{2/3}/\omega_n$, where $\omega_n \to \infty$ arbitrarily slowly. That is, almost every graph has the property that the biggest tree that is a component has order about $n^{2/3}$.

Evidently there are no *other* components of order $\geq n^{2/3}\omega_n$ either.

For $0 < c < 1/2$, we return to the expression in (4.2.8) for $E(X_m)$ and note that $0 < 2ce^{1-2c} < 1$ for $c > 0$ and $c \neq 1/2$. So, following Erdös and Rényi, set

(4.2.12) $$e^{-\alpha} = 2ce^{1-2c}$$

to define α, and from (4.2.8) obtain the bound

(4.2.13) $$E(X_m) = O(1)\frac{n}{m^{5/2}}e^{-\alpha m}.$$

Now a nice choice for m can be made:

(4.2.14) $$m = \frac{1}{\alpha}\left(\log n - \frac{5}{2}\log\log n\right) + \omega_n$$

where $\omega_n \to \infty$ arbitrarily slowly.

Combining (4.2.13) and (4.2.14) we find

(4.2.15) $$E(X_m) = O(1)e^{-\alpha\omega_n} = o(1),$$

which proves the first part (4.2.24) of Theorem 4.2.2 for $c < 1/2$. The *same* *choice* in (4.2.12) for m works for $c > 1/2$, but it is not quite as easy to establish, because (4.2.8) is no longer applicable. A much better approximation of $(n)_k/n^k$ than (4.2.5) must be used.

So we back up to formula (4.2.4) and split the sum on the right side as follows. First we consider the range of k such that $k^2 = O(n)$. Then we will still have (4.2.7) and hence (4.2.8). Then the same argument used earlier shows that the lower portion of the sum in (4.2.4) goes to zero. As for the upper portion, we are left with the task of proving that for some constant K

(4.2.16) $$n\Sigma\frac{(n)_k}{n^k}\frac{1}{k^{5/2}}(e^{ck/n-\alpha})^k = o(1),$$

where the sum is over all k with $Kn^{1/2} \leq k \leq n$. Call the left side of (4.2.16) $F(n)$, and replace the term $k^{5/2}$ by its smallest value so that we have

(4.2.17) $$F(n) = O(1)n^{-1/4}\sum\frac{(n)_k}{n^k}(e^{ck/n-\alpha})^k.$$

Now we use the following bound for $(n)_k/n^k$, whose verification is left as exercise 4.2.3:

(4.2.18) $$\frac{(n)_k}{n^k} \leq \exp\left\{ -\frac{(k-1)^2}{n} \sum_{i=1}^{\infty} \frac{1}{i(i+1)} \left(\frac{k-1}{n}\right)^{i-1}\right\}.$$

Using this bound we can write

(4.2.19) $$F(n) = O(1)n^{-1/4}\Sigma\left[\exp\left\{f\left(\frac{k-1}{n}, c\right)\right\}\right]^{k-1},$$

where

(4.2.20) $$f(x,c) = -\alpha + cx - \sum_{i=1}^{\infty} \frac{x^i}{i(i+1)}.$$

To complete the task of establishing formula (4.2.16), we just need to show that for some $\varepsilon > 0$

(4.2.21) $$\exp\{f(x,c)\} \leq 1 - \varepsilon$$

for all x, c with $c > 1/2$ and $0 < x < 1$. We first simplify the right side of (4.2.20) with the following identity (see exercise 4.2.4):

(4.2.22) $$\sum_{i=1}^{\infty} \frac{x^i}{i(i+1)} = \frac{(1-x)\log(1-x) + x}{x}.$$

Therefore we have from (4.2.12), (4.2.20) and (4.2.22):

(4.2.23) $$f(x,c) = (\log 2c) - 2c + cx - \frac{1-x}{x}\log(1-x).$$

Now we can look for the critical points of $f(x,c)$. On setting the partial derivatives of f with respect to x and c equal to zero, we find that the only solution is $x = 0$ and $c = 1/2$, where $f(0, 1/2) = 0$. Therefore the largest value of f occurs on the boundary of the region $0 \leq x \leq 1$, $c > 0$ and, indeed, is at $(0, 1/2)$. The details are left to exercise 4.2.5.

Thus we have (4.2.21), and so the sum on the right side of (4.2.19) converges. This, in turn, implies (4.2.16). Note that our proof of (4.2.24) for $c > 1/2$ works for all $c > 0$, $c \neq 1/2$.

The second-moment method can again be used, and the consequences of these efforts are summarized below.

Theorem 4.2.2.   In Model A with $p = 2c/n$, $c \neq 1/2$,

(4.2.24) $$P(X_m \geq 1) \to 0$$

for

$$m = \frac{1}{\alpha}\left(\log n - \frac{5}{2}\log\log n\right) + \omega_n,$$

but

(4.2.25) $$P(X_m \geq 1) \to 1$$

for

$$m = \frac{1}{\alpha}\left(\log n - \frac{5}{2}\log\log n\right) - \omega_n,$$

where $\omega_n \to \infty$ arbitrarily slowly. That is, in almost every graph the largest component that is a tree has order about $\log n$.

Perhaps the most startling development is the appearance of the giant* component with $c > 1/2$. To see how this comes about, let $X(G)$ be the order of the biggest component of G. We should try to show that almost all graphs have $X \geq (1 - \varepsilon)n$ for small $\varepsilon > 0$. Therefore we look for a nice bound on $P(X \leq (1 - \varepsilon)n)$. A graph G that satisfies the condition

(4.2.26) $$X(G) \leq (1 - \varepsilon)n$$

can have its vertices partitioned in two nonempty sets of m_1 and m_2 vertices, with no edge having a vertex in each of the two sets. Therefore the sum of the probabilities of all graphs that can be partitioned in this way is bounded as follows:

(4.2.27) $$\Sigma P(G) \leq (1 - p)^{m_1 m_2}.$$

Next we show that each graph can be partitioned in such a way that the product $m_1 m_2$ satisfies

(4.2.28) $$m_1 m_2 \geq \frac{\varepsilon}{2}\left(1 - \frac{\varepsilon}{2}\right)n^2.$$

To do this, Erdös and Rényi used the following nifty lemma whose proof is left as an exercise.

Lemma. Let $a_1, \ldots, a_r$ be positive numbers whose sum is 1. If $\max_{1 \leq i \leq r} a_i \leq (1 - \varepsilon)$, then there is a number k between 1 and $r - 1$ such that

(4.2.29) $$\frac{\varepsilon}{2} \leq \sum_{i=1}^{k} a_i \leq 1 - \frac{\varepsilon}{2}$$

---

*The group theorists have their monster, but we have the whopper.

and

$$(4.2.30) \qquad \frac{\varepsilon}{2} \le \sum_{i=k+1}^{r} a_i \le 1 - \frac{\varepsilon}{2}.$$

Now just let the r components of the graph have order $a_i n$ for $i = 1$ to r. Then the lemma implies that m_1 and m_2 can be chosen such that

$$(4.2.31) \qquad \frac{\varepsilon}{2} \le \frac{m_1}{n}, \frac{m_2}{n} \le 1 - \frac{\varepsilon}{2}.$$

By definition, $m_1 + m_2 = n$, so the product $m_1 m_2$ is minimized when $m_1 = n\varepsilon/2$ and $m_2 = n(1 - \varepsilon/2)$, and so (4.2.28) is verified.

Since $0 < \varepsilon < 1$, we also have $1 - \varepsilon/2 > 1/2$, and (4.2.28) can be simplified:

$$(4.2.32) \qquad m_1 m_2 \ge \frac{\varepsilon n^2}{4}.$$

Combining this inequality with (4.2.27) and the crude upper bound 2^n for the number of partitions of the n vertices, we have

$$(4.2.33) \qquad P(X \le (1 - \varepsilon)n) \le 2^n (1 - p)^{\varepsilon n^2/4}.$$

From (4.2.33) with $p = 2c/n$, we have

$$(4.2.34) \qquad P(X \le (1 - \varepsilon)n) = O(1) \left(\frac{2}{e^{c\varepsilon/2}} \right)^n,$$

and so this probability has limit zero provided

$$(4.2.35) \qquad 2 \log 2 < c\varepsilon.$$

Thus if c and ε satisfy (4.2.35) and $\varepsilon < 1/2$, then almost every graph has a unique giant component of order $> n/2$.

This rough result is substantially refined in the following theorem of Erdös and Rényi.

Theorem 4.2.3. Let $X(G)$ denote the order of the largest component of G. In Model A, with $p = 2c/n$, $c > 1/2$ for any $\varepsilon > 0$, we have

$$(4.2.36) \qquad P\left(\left| \frac{X}{n} - G(c) \right| < \varepsilon \right) \to 1$$

where

(4.2.37)
$$G(c) = 1 - \frac{x(c)}{2c}$$

and

(4.2.38)
$$x(c) = \sum_{k=1}^{\infty} \frac{k^{k-1}}{k!} \left(2ce^{-2c}\right)^k.$$

That is, almost all graphs have about $G(c)n$ vertices in their largest component.

The infinite series on the right side of (4.2.38) that defines $x(c)$ converges whenever $2ce^{-2c} \leq e^{-1}$. Therefore (4.2.38) defines $x = x(c)$ for all $c > 0$. Note that $x(c)/2c \rightarrow 0$ as $c \rightarrow +\infty$, and hence $G(c) \rightarrow 1$. The ratio $x(c)/2c$ is, in fact, the asymptotic proportion of vertices in acyclic components!

Theorem 4.2.4.   Let $X(G)$ denote the number of vertices of G that belong to components that are trees. Then in Model A with $p = 2c/n$ and constant $c > 0$, we have

(4.2.39)
$$\frac{E(X)}{n} \rightarrow \begin{cases} 1 & \text{for } c \leq 1/2 \\ \dfrac{x(c)}{2c} & \text{for } c > 1/2. \end{cases}$$

Proof.   In the proof we will make use of a simple observation about the function $x = x(c)$ that follows from some basic facts established in enumeration theory (see Section 1.7 of [HP73] or Section 4.2 of [M70]). For $c > 0$, the function $x = x(c)$ is the only root of the equation

(4.2.40)
$$xe^{-x} = 2ce^{-2c}$$

that satisfies

(4.2.41)
$$0 < x(c) \leq 1.$$

Of course, if $0 < c \leq 1/2$, then $x(c)$ must be $2c$, and so in this case

(4.2.42)
$$2c = \sum_{k=1}^{\infty} \frac{k^{k-1}}{k!} \left(2ce^{-2c}\right)^k.$$

Now we state the expected value of the random variable X defined in the hypothesis:

$$(4.2.43) \quad E(X) = \sum_{k=1}^{n} k \binom{n}{k} k^{k-2} p^{k-1} (1-p)^{\binom{k}{2} - (k-1) + k(n-k)}.$$

Note the similarity of the summands here to those in (4.2.1). The only difference is the first factor k present in (4.2.43). We will shortly make use of our analysis of (4.2.1).

With $p = 2c/n$ and k fixed, it is easy to estimate the kth summand of (4.2.43) and find

$$(4.2.44) \quad \binom{n}{k} k^{k-1} p^{k-1} (1-p)^{\binom{k}{2} - (k-1) + k(n-k)} \sim \frac{n}{2c} \frac{k^{k-1}}{k!} (2ce^{-2c})^k.$$

Therefore if we divide both sides of (4.2.43) by n, we find

$$(4.2.45) \qquad \liminf_{n \to \infty} \frac{E(X)}{n} \geq \frac{1}{2c} \sum_{k=1}^{t} \frac{k^{k-1}}{k!} (2ce^{-2c})^k$$

for any $t \geq 1$. Therefore

$$(4.2.46) \qquad \liminf_{n \to \infty} \frac{E(X)}{n} \geq \frac{1}{2c} \sum_{k=1}^{\infty} \frac{k^{k-1}}{k!} (2ce^{-2c})^k.$$

But if $0 < c \leq 1/2$, by (4.2.42) the right side of (4.2.46) is exactly 1. On the other hand, by definition, $E(X) \leq n$, and therefore $E(X)/n \to 1$.

Now we consider the second part of the theorem where $c > 1/2$. After division by n, the formula (4.2.43) for $E(X)$ is split into two parts. The lower part consists of all terms for which $k \leq m - 1$, where m is defined by (4.2.14). The upper part is summed over all k with $m \leq k \leq n$. But this part differs from $E(X_m)$ in (4.2.1) only by its extra factor k/n which is ≤ 1. Therefore $E(X_m)$ is an upper bound for this part, and we have already shown that $E(X_m) \to 0$. The range of k for the lower part implies that $k^2 = o(n)$, and so the summands can be substantially simplified. In particular, we can use

$$(4.2.47) \qquad (1-p)^{\binom{k}{2} - (k-1) + k(n-k)} \sim (e^{-2c})^k$$

and formula (3.5) of Appendix III to show that

$$(4.2.48) \qquad \frac{E(X)}{n} = o(1) + \frac{1}{2c} \sum_{k=1}^{m-1} \frac{k^{k-1}}{k!} (2ce^{-2c})^k.$$

The second part of (4.2.39) now follows from (4.2.48).   □

With just a bit more effort the result of (4.2.39) for $c < 1/2$ can be strengthened considerably. First, for $k^2 = o(n)$ we will use

$$(4.2.49) \qquad \frac{(n)_k}{n^k}(1 - p)^{\binom{k}{2} - (k-1) + k(n-k)} = e^{-2ck}\left[1 + O\left(\frac{k^2}{n}\right)\right].$$

Then, using the same technique to establish (4.2.13), we can show for $m = (\log n)/\alpha$ that

$$(4.2.50) \quad E(X) = \frac{n}{2c}\sum_{k=1}^{m}\frac{k^{k-1}}{k!}(2ce^{-2c})^k\left[1 + O\left(\frac{k^2}{n}\right)\right] + \frac{O(1)}{m^{3/2}}.$$

Since $c < 1/2$, it follows from (4.2.42) and (4.2.50) that

$$(4.2.51) \qquad\qquad E(x) = n - O(1).$$

This formula for $E(X)$ can be used to estimate the number of vertices in acyclic components of almost all graphs. To do this we define a new random variable

$$(4.2.52) \qquad\qquad Y = n - X,$$

which counts the vertices that are not in acyclic components. Now we apply Markov's inequality (Proposition 5.2 of Appendix V) with $t = \omega_n$, which tends to infinity arbitrarily slowly. The result is

$$(4.2.53) \qquad\qquad P(n - X \geq \omega_n) \leq \frac{E(n - X)}{\omega_n}.$$

The next theorem follows from (4.2.51) and the linearity of the expectation.

Theorem 4.2.5. Let $X(G)$ be the number of vertices of G that belong to components that are trees. In Model A with $p = 2c/n$ and $c < 1/2$,

$$(4.2.54) \qquad\qquad P(n - X \leq \omega_n) \to 1,$$

where $\omega_n \to \infty$ arbitrarily slowly. That is, almost all graphs have fewer than ω_n vertices in the components with cycles.

We have seen in Theorem 4.1.1 that the limiting probability of a cycle is not zero when $c < 1/2$. Erdös and Rényi found, however, that almost all graphs have no components with more than one cycle!

Theorem 4.2.6. In Model A with $p = 2c/n$ and $c < 1/2$, almost all graphs have no components with more than one cycle, that is, the components are trees or are unicyclic.

Proof. Let $X(G)$ be the number of vertices in components of order at most ω_n that have more edges than vertices. It follows from Theorem 4.2.5 that we need only find a slowly growing function ω_n such that $E(X) \to 0$. Using the crude upper bound $p^{k+1}2^{\binom{k}{2}}$ for the probability that a graph of order k has more than k edges, we have

$$(4.2.55) \qquad E(X) \le \sum_{k=1}^{\omega_n} k\binom{n}{k} p^{k+1}2^{\binom{k}{2}}(1-p)^{k(n-k)}.$$

If $k^2 = o(n)$, then

$$(4.2.56) \qquad (1-p)^{k(n-k)} \sim e^{-2ck},$$

and so

$$(4.2.57) \qquad E(X) = O(1)\frac{1}{n}\sum_{k=1}^{\omega_n} 2^{k^2/2}(2ce^{-2c})^k$$

$$= O(1/n)2^{(\omega_n)^2/2}.$$

Therefore if we just take $\omega_n = \sqrt{\log n}$, we will have $E(X) \to 0$ as required. $\qquad\square$

Since trees and unicyclic graphs are planar, if $p = 2c/n$ with $c < 1/2$, we also know that almost all graphs are planar. We conclude this section with yet another remarkable property of the double-jump threshold $p = 2c/n$. It is also a threshold for planarity. We now describe the main steps used by Erdös and Rényi to establish this important fact of graphical evolution.

Our aim is to show that if $c > 1/2$, almost every graph contains a subgraph homeomorphic to $K_{3,3}$. Note that $K_{3,3}$ is isomorphic to a cycle of order 6 whose vertices are colored alternately green and white and that has three more diagonal edges joining vertices of different colors. However, we will not be able to show that almost every graph contains a subgraph of fixed order homeomorphic to $K_{3,3}$. Such a subgraph has $k \ge 6$ vertices and $l = k + 3$ edges. Indeed, if $k \ge 4$ and $l \ge k + 1$,

$$(4.2.58) \qquad pn^{k/l} = (2c)n^{-(l-k)/l} \to 0,$$

and it follows from Theorem 3.1.2 that almost no graph contains a subgraph of fixed order that is homeomorphic to $K_{3,3}$. But if we let the order vary, the story is quite different. Therefore let $X(G)$ be the number of cycles of order k with exactly d chords or diagonals. Then the expectation is

$$
(4.2.59) \quad
\begin{aligned}
E(X) &= \sum_{k=4}^{n} \binom{n}{k} \frac{(k-1)!}{2} \left(\binom{\binom{k}{2} - k}{d} \right) p^{k+d} (1-p)^{\binom{k}{2} - k - d} \\
&= O(1) \sum_{k=4}^{n} \frac{(n)_k}{n^k} \frac{(2c)^k}{k} \left(\frac{k^2}{n} \right)^d e^{-ck^2/n},
\end{aligned}
$$

where the big-O term depends on c and d. If $c < 1/2$, then $E(X) = O(n^{-d})$ and so $E(X) \to 0$, which is not unexpected, because a cycle with one diagonal is forbidden by Theorem 4.2.6. But if $c > 1/2$, just the term in the sum for $k = \sqrt{n}$ tends to infinity and so $E(X) \to \infty$. Erdös and Rényi found that if the constant c is replaced by

$$
(4.2.60) \qquad\qquad 2c = 1 + \frac{\lambda}{\sqrt{n}},
$$

where λ is any fixed real number, then $E(X)$ can be represented by an integral

$$
(4.2.61) \qquad\qquad E(X) = O(1) \int_0^\infty y^{2d-1} e^{\lambda y - y^2} \, dy.
$$

Furthermore they discovered that this distribution obeyed Poisson's law. Hence, if the constant c is greater than $1/2$, almost every graph is sure to have cycles with three diagonal edges. There are $\binom{k}{2} - k$ possible diagonal edges in a cycle of order k. Therefore there are

$$
\left(\binom{\binom{k}{2} - k}{3} \right)
$$

ways to select $d = 3$ of these, and of these $\binom{k}{6}$ will result in subgraphs homeomorphic to $K_{3,3}$. For large k the probability is approximately $3! 2^3 / 6! = 1/15$ that a nonplanar subgraph will be formed. Therefore the expected number of these is about $1/15$ of $E(X)$ in (4.2.59) when $d = 3$. Therefore if $c > 1/2$, $E(X)/15 \to \infty$, and the same argument used above for $E(X)$ shows that almost every graph has one of these nonplanar subgraphs formed by a cycle with three diagonal edges.

Theorem 4.2.7. In Model A with $p = 2c/n$, if $c < 1/2$, almost all graphs are planar, but if $c > 1/2$, almost all graphs are nonplanar.

The problem of determining the probability of planarity for almost all graphs when $p \sim 1/n$ is apparently still open. Erdös and Rényi were able to show that if $p = (1 + \lambda/\sqrt{n})/n$, then the probability of being nonplanar has a positive lower limit, but they did not calculate its value. They observed that the value could even be 1, but is seems unlikely that almost all graphs are nonplanar when $c = 1/2$.

Exercises 4.2

1. Show that in Model A with $p = 1/n$ and any $\varepsilon > 0$, almost all graphs have no components of order $> (1 + \varepsilon)n/2$.

2. Verify formula (4.2.2).

3. Verify the upper bound for $(n)_k/n^k$ in formula (4.2.18).

4. Verify the identity of formula (4.2.22).

5. Compute the partials of f in (4.2.23) with respect to x and c, set them equal to zero, and show that $(x, c) = (0, 1/2)$ is an absolute maximum of $f(x, c)$ for $0 \le x \le 1$, $c > 0$.

6. Prove the lemma in this section.

4.3.   CONNECTIVITY AND BEYOND

When $q \sim cn$ or $p = 2c/n$ with $c > 1/2$, there is a unique giant component of order about $G(c)n$, and the remaining vertices, about $[x(c)/2c]n$, belong to components that are trees of order at most $\log n$. As c increases, $G(c) \to 1$ and $x(c)/2c \to 0$, so the giant grows by absorbing these trees. The larger the tree, the more likely it is to be consumed by the giant.

We know from Section 3.1 that even with many more edges as when $q \sim (c/2)n \log n$ or $p = c(\log n)/n$ with $0 < c < 1$, there are still sure to be isolated vertices. These niblets must be the last of the trees to be scooped up by the giant on the way to connectivity. We also know that when $c > 1$, there are no more isolated vertices. So it is reasonable to assume that for $c > 1$ and $p = c(\log n)/n$, all small components (munchies) have been captured by the giant and full connectivity has been achieved. Thus we should try to show that for $c > 1$ there are no components of order $< n$. Let $X(G)$ be the number of these components; then

$$(4.3.1) \qquad E(X) = \sum_{k=1}^{n-1} \binom{n}{k} P_k (1 - p)^{k(n-k)},$$

where P_k is the sum of the probabilities of all connected graphs of order k.

Since $P_k \leq 1$,

(4.3.2)
$$E(X) \leq \sum_{k=1}^{n-1} \binom{n}{k}(1-p)^{k(n-k)}.$$

Since the sum in (4.3.2) is symmetric, we can write

(4.3.3)
$$E(X) \leq 2 \sum_{k=1}^{\lfloor n/2 \rfloor} \binom{n}{k}(1-p)^{k(n-k)}.$$

We have already seen in (3.1.21) that the first term in the sum is the expected number of isolated vertices and

(4.3.4)
$$\binom{n}{1}(1-p)^{1(n-1)} \sim n^{1-c}.$$

The best we can hope for, then, is to show that $E(X) = O(n^{1-c})$. Therefore we will split the rest of the sum into two parts according as $k \leq$ or $> \lfloor \varepsilon n/2 \rfloor$, and ε, between 0 and 1, will be chosen later.

Now we use

$$\binom{n}{k} \leq n^k$$

$$(1-p)^n \leq n^{-c}$$

and, for $2 \leq k \leq \lfloor \varepsilon n/2 \rfloor$,

$$n - k \geq n\left(1 - \frac{\varepsilon}{2}\right)$$

to obtain

$$\binom{n}{k}(1-p)^{k(n-k)} \leq \left(n^{1-c(1-\varepsilon/2)}\right)^k.$$

Note that since $c > 1$, ε can be chosen so small that

$$\alpha = c\left(1 - \frac{\varepsilon}{2}\right) - 1 > 0,$$

that is, $\varepsilon < 2(1 - 1/c)$ will do. Then

(4.3.5)
$$\begin{aligned}
\sum_{k=2}^{\lfloor \varepsilon n/2 \rfloor} \binom{n}{k}(1-p)^{k(n-k)} &\leq \sum_{k=2}^{\lfloor \varepsilon n/2 \rfloor} \left(\frac{1}{n^\alpha}\right)^k \\
&\leq \left(\frac{1}{n^\alpha}\right)^2 \sum_{k=0}^{\infty} \left(\frac{1}{n^\alpha}\right)^k \\
&= \frac{1}{n^{2\alpha}} O(1) \\
&= O(n^{1-c}),
\end{aligned}$$

where the last step requires $\varepsilon < (1 - 1/c)$.

The upper portion of the sum where $\lfloor \varepsilon n/2 \rfloor < k \le \lfloor n/2 \rfloor$ is even easier. Here we use the crude bound

$$\binom{n}{k} \le 2^n$$

and

$$k(n - k) \ge \frac{\varepsilon n}{2}\left(\frac{n}{2}\right),$$

and since the length of the sum is $\le n$ we have

(4.3.6)
$$\sum_{k > \lfloor \varepsilon n/2 \rfloor}^{\lfloor n/2 \rfloor} \binom{n}{k}(1 - p)^{k(n-k)} \le n2^n(1 - p)^{\varepsilon n^2/4}$$

$$\le n2^n n^{-c\varepsilon n/4}$$

$$= O(n^{1-c})$$

where the last step only requires c and ε to be positive.

These results are now summarized.

Theorem 4.3.1. Let $\mathscr{C}$ be the set of connected graphs of order n. In Model A with $p = c(\log n)/n$, the expected number of components of order $\le n - 1$ is $O(n^{1-c})$; hence $P(\overline{\mathscr{C}}) = O(n^{1-c})$. Therefore for $c > 1$:

$$P(\mathscr{C}) \to 1,$$

that is, almost every graph is connected.

Now we know that the threshold function of Theorem 3.1.1 for isolated vertices also works for connectivity. But we do not know what the limiting probability of connectivity is when $p \sim (\log n)/n$. However, this latter probability is high enough that we can be certain there is a giant component and possibly some small components outside the giant. Since we are so close to connectivity, there is a possibility that *all* the components outside the giant are isolated vertices. If this is indeed the case, then the probability of connectivity in the limit should be the same as the probability of no isolated vertices, already worked out in Theorem 3.2.1. Therefore to find a sharp threshold for connectivity we will consider the sharp threshold in (3.2.10) for isolated vertices and show that the components outside the giant are just isolated vertices.

Theorem 4.3.2. Let $\mathscr{A}$ be the set of all graphs of order n that have one component of order at least 2, with all other components isolated vertices. In

Model A with $p = (\log n)/n + x/n$,

$$P(\mathscr{A}) \to 1,$$

that is, almost all graphs consist of a giant component plus, perhaps, isolated vertices.

Proof. We will show that $P(\bar{\mathscr{A}}) \to 0$, and we begin by splitting the probability of this event into two parts depending on the order of the biggest component.

Let M be a number that depends on n and will be chosen later. Then $\mathscr{E}_M$ is the set of graphs of order n in which the biggest component has at least $n - M$ vertices. We want

(i) $P(\bar{\mathscr{A}} \cap \mathscr{E}_M) \to 0$

and

(ii) $P(\bar{\mathscr{A}} \cap \bar{\mathscr{E}}_M) \to 0$.

First we concentrate on (i) and consider any graph G in $\bar{\mathscr{A}} \cap \mathscr{E}_M$. Suppose this graph has k vertices outside a biggest component. Then, since $G \in \mathscr{E}_M$, $n - k \geq n - M$ and so $k \leq M$. Also since $G \in \bar{\mathscr{A}}$, we must have $k \geq 2$, and there is at least one edge in the subgraph induced by these k vertices. Therefore

(4.3.7) $$P(\bar{\mathscr{A}} \cap \mathscr{E}_M) \leq \sum_{k=2}^{M} \binom{n}{k}(1 - p)^{k(n-k)} p2^{\binom{k}{2}},$$

where $p2^{\binom{k}{2}}$ is just a crude upper bound for the probability that a graph of order k has at least one edge.

Now we turn our attention to condition (ii). In this case, a graph G in $\bar{\mathscr{A}} \cap \bar{\mathscr{E}}_M$ has $M < k$, which is again the number of vertices outside a biggest component of G.

Note that k cannot be too large, otherwise all the components will be so small that there will not be enough room to fit the expected number of edges. From exercise 3.1.2 and equation (3.2.10) it follows that almost all graphs have at least $[(1 - \varepsilon)/2]n \log n$ edges. Asking how large k can be is the same as asking how small $n - k$, the order of a biggest component, can be. Now $n - k$ is smallest for a graph in which each component has about $n - k$ vertices and is fully packed with edges. Therefore we must have

(4.3.8) $$\frac{n}{n - k}\binom{n - k}{2} \geq \frac{(1 - \varepsilon)}{2} n \log n$$

for almost all graphs. From which it follows that

$$(4.3.9) \qquad k \leq n - (1 - \varepsilon)\log n.$$

Now we can write

$$(4.3.10) \qquad P(\bar{\mathscr{A}} \cap \bar{\mathscr{E}}_M) \leq P(\bar{\mathscr{E}}_M) \leq \sum_{k > M} \binom{n}{k}(1 - p)^{k(n-k)},$$

where the upper bound in the sum is given by (4.3.9). Now it is time to determine M as a function of n so that the bounds in (4.3.7) and (4.3.10) will both go to zero as $n \to \infty$. Here is a rough outline of the rest of the proof.

Looking at the right side of (4.3.7) we first estimate

$$(4.3.11) \qquad (1 - p)^{k(n-k)} \sim \left(\frac{e^{-x}}{n}\right)^k,$$

provided $k = o(\sqrt{n/\log n})$. Then using $\binom{n}{k} \leq n^k/k!$, we have

$$(4.3.12) \qquad P(\bar{\mathscr{A}} \cap \mathscr{E}_M) \leq p \sum_{k=2}^{M} 2^{\binom{k}{2}} \frac{(e^{-x})^k}{k!} = p2^{M^2/2}O(1),$$

and it is easy to see that this last expression goes to zero for $M = \lfloor \log \log n \rfloor$.

Now we consider the right side of (4.3.10) and observe that

$$(4.3.13) \qquad \overline{P}(\bar{\mathscr{A}} \cap \bar{\mathscr{E}}_M) \leq 2 \sum_{k=M}^{\lfloor n/2 \rfloor} \binom{n}{k}(1 - p)^{k(n-k)}.$$

If $k \leq n/2$, then $n - k \geq n/2$, and so

$$(4.3.14) \qquad \begin{aligned} (1 - p)^{k(n-k)} &\leq (1 - p)^{kn/2} \\ &\leq (e^{-np})^{k/2} \\ &\leq \left(\frac{1}{n^{1/2}e^{x/2}}\right)^k. \end{aligned}$$

Combining this inequality with

$$(4.3.15) \qquad \binom{n}{k} \leq \left(\frac{en}{k}\right)^k,$$

we have for $k \leq n/2$

$$(4.3.16) \qquad \binom{n}{k}(1 - p)^{k(n-k)} \leq \left(\frac{n^{1/2}}{k}\frac{e}{e^{x/2}}\right)^k.$$

Therefore that portion of the sum in (4.3.13) for which k is a bit bigger than $n^{1/2}$ will go to zero. Specifically, if we choose the constant a big enough that

(4.3.17)
$$\frac{e}{ae^{x/2}} = \alpha < 1,$$

then

(4.3.18)
$$\sum_{k \geq a\sqrt{n}}^{\lfloor n/2 \rfloor} \binom{n}{k}(1 - p)^{k(n-k)} \leq \sum_{k=a\sqrt{n}}^{\infty} \alpha^k$$
$$= O(\alpha^{a\sqrt{n}})$$
$$= o(1).$$

The rest of the sum is handled as follows. First note that

(4.3.19)
$$(1 - p)^{k(n-k)} \leq \left(\frac{1}{e^{(n-k)p}}\right)^k.$$

But if $k \leq a\sqrt{n}$, on substitution for k and p we also have

(4.3.20)
$$(1 - p)^{k(n-k)} = O\left(\frac{1}{n}\right)^k.$$

Therefore for some constant b and n sufficiently large,

(4.3.21)
$$\binom{n}{k}(1 - p)^{k(n-k)} \leq \frac{b^k}{k!}.$$

So the other portion of the sum in (4.3.13) is bounded as follows:

(4.3.22)
$$\sum_{k=M}^{a\sqrt{n}} \binom{n}{k}(1 - p)^{k(n-k)} \leq \sum_{k=M}^{a\sqrt{n}} \frac{b^k}{k!} = o(1). \qquad \square$$

Now the sharp threshold function for connectivity is seen to be the same as that for isolated vertices.

Corollary 4.3.1. Let $\mathscr{C}$ be the set of connected graphs of order n. In Model A with $p = (\log n + x)/n$, the probability of connectivity has limit

(4.3.23)
$$P(\mathscr{C}) \to e^{-e^{-x}}.$$

Proof. Let $\mathscr{B}$ be the set of graphs, of order n with no isolated vertices. Recall that the limiting probability of $\mathscr{B}$ was worked out in Theorem 3.2.1.

Then

(4.3.24) $$\mathscr{C} = \mathscr{B} \cap \mathscr{A},$$

and so

(4.3.25) $$P(\mathscr{B}) = P(\mathscr{C}) + P(\mathscr{B} \cap \overline{\mathscr{C}}).$$

But $\mathscr{B} \cap \overline{\mathscr{C}} \subseteq \mathscr{A}$, and therefore by the theorem $P(\mathscr{B} \cap \overline{\mathscr{C}}) \to 0$. Now we have $P(\mathscr{B}) = P(\mathscr{C}) + o(1)$, but since $P(\mathscr{B}) \nrightarrow 0$, this implies $P(\mathscr{B}) \sim P(\mathscr{C})$. □

The results above are, of course, the work of Erdös and Rényi, but they are found in the early paper [ErR59] and expressed in the terms of Model B. There is no particular difficulty in generalizing the corollary for certain types of connectivity in uniform hypergraphs [PaR-U]. The overall approach can also be successfully applied to the study of connectivity of the deleted neighborhoods of the vertices [ErPR83].

Wright [W70a] has obtained good asymptotic estimates for the number $C_{n,q}$ of connected graphs when the number q of edges is well beyond the threshold of Corollary 4.3.1. His results apply, for example, when $q = cn \log n$ with $c > 1/2$, and his methods are quite different since they are based on a recurrence relation. An important consequence is a good approximation of the number of disconnected graphs.

Almost every graph is not only connected when $p = c(\log n)/n$, $c > 1$, but is also hamiltonian, that is, it possesses a spanning cycle. An early result in this direction was found by Wright [W74b], who showed that in Model B with $q = n^{3/2}\omega_n$, $\omega_n \to \infty$ arbitrarily slowly, almost every graph is hamiltonian. The most important breakthrough was made by Pósa [Po76], who used Model A with $p = c(\log n)/n$ to show that almost all graphs are hamiltonian when c is somewhat larger than 1. A proof for $c > 9$ can be found in Bollobás's book [Bo79]. Koršunov evidently found a fast probabilistic algorithm that produces a spanning cycle for almost every graph when $c > 1$. He announced in [Ko76] that in Model B almost all graphs are hamiltonian iff

(4.3.26) $$q \sim \frac{n}{2}[\log n + \log \log n + \omega_n]$$

with $\omega_n \to \infty$. This latter threshold just guarantees that almost every graph has minimum degree at least 2 [compare (5.1.8), with $d = 2$].

Komlós and Szemerédi [KoS83] have just established for hamiltonian graphs a threshold result of the form of Corollary 4.3.1.

Theorem 4.3.3. Let $\mathscr{H}$ be the set of hamiltonian graphs of order n. In Model A with

$$(4.3.27) \qquad p = \frac{\log n + \log \log n + 2c_n}{n},$$

the probability of hamiltonicity has the limit

$$(4.3.28) \qquad P(\mathscr{H}) \to \begin{cases} 0 & \text{if } c_n \to -\infty \\ e^{-e^{-2c}} & \text{if } c_n \to c \\ 1 & \text{if } c_n \to +\infty \end{cases}.$$

Wright [W83] has also just discovered a very nice estimate of the number $H_{n,q}$ of labeled hamiltonian graphs of order n and size q for very small q.

Theorem 4.3.4. If $q - n = o(n)$, then

$$(4.3.29) \qquad \frac{H_{n,q}}{\dfrac{(n-1)!}{2}\left(\dbinom{\binom{n}{2}-n}{q-n}\right)} \to 1.$$

The evolution of unlabeled graphs has been studied in some detail by Wright. The relevant Wright papers to read include [W70b], [W72a], [W74a], [W74c], [W75a], [W76a], [W76b] and [W76c]. He observed [W76a] that the evolution of unlabeled graphs is relatively simple but the proofs are more complicated because they require the formula for $g_{n,q}$ from Pólya's enumeration theorem (see Chapter 4 of [HP73]).

An important advance [W70b] occurred when Wright was able to prove that (1.1.6) is a necessary and sufficient condition for (1.1.5). When (1.1.6) holds, properties of almost all labeled graphs are also held by almost all unlabeled graphs. For example, if $q \sim cn \log n$ and $c > 1/2$, then almost all labeled graphs are connected. This is just Theorem 4.3.1 in Model B. But (1.1.6) is satisfied, and so we have (1.1.5). Let $c_{n,q}$ denote the number of unlabeled connected graphs of order n and size q. It now follows from (1.1.1) and (1.1.5) that

$$(4.3.30) \qquad \frac{c_{n,q}}{g_{n,q}} \sim 1,$$

that is, almost all unlabeled graphs are connected. Therefore much asymptotic enumeration for unlabeled graphs reduces to that of labeled graphs. On the other hand, Wright found the situation far from dull when (1.1.6) is not satisfied [W76b]. Note that this is exactly the stage of evolution at which connectivity occurs. In fact, there is quite a drastic difference between the labeled random graph and the unlabeled as q increases up to this threshold. For example, the probability distribution function for connectivity is completely different. If $q \sim n(\log n + x)/2$, Wright* [W75a] proved

$$(4.3.31) \qquad \frac{c_{n,q}}{g_{n,q}} \to 1 - e^{-x}.$$

For a good introduction to this subject, the reader should consult the two papers [W76a] and [W76b].

SUMMARY

Model A	Model B	Description of Almost All Graphs
$p = 2c/n$	$q \sim cn$	
	$0 < c < 1/2$	All components are trees or unicyclic. The largest component is a tree of order about $\log n$.
	$c = 1/2$	There are certain to be cycles, and the largest component has order about $n^{2/3}$.
	$c > 1/2$	There is a unique giant component of order about $G(c)n$. All but $o(n)$ vertices belong to the giant or trees of order at most $\log n$.
$p = c\dfrac{\log n}{n}$	$q \sim c\frac{1}{2}n \log n$	
	$0 < c < 1$	The random graph is disconnected.
	$c > 1$	The graph is not only connected but also hamiltonian.
$p = \dfrac{\log n}{n} + \dfrac{x}{n}$	$q \sim \frac{1}{2}(n \log n + xn)$	Outside the giant there are only isolated vertices, and the probability of connectivity is $\exp\{-\exp[-x]\}$.

*Sir Edward was knighted not for his work in number theory, his celebrated book written with G. H. Hardy nor for his many important papers in enumerative graph theory, but for his outstanding contributions as an administrator at the University of Aberdeen, where he served many years as Vice Chancellor.

Exercises 4.3

1. Prove that (1.1.4) implies that almost every graph has the identity group in Model A with $p = 1/2$. What is the corresponding result for (1.1.5)?

2. Use exercise 4.3.1 and Theorem 3.1.1 to prove that (1.1.5) implies (1.1.6), that is, the easy half of Wright's theorem.

3. If $p = c(\log n)/n$ and $c > 1$, show that

$$P(\bar{\mathscr{A}}) = O\left\{ \frac{(\log n)2^{(\log\log n)^2/2}}{n^{2c-1}} \right\}.$$

4. Use Wright's result on hamiltonicity in Model B (see text) to show that if the probability of an edge is fixed, then almost all graphs are hamiltonian in Model A.

5

SELECTED HIGHLIGHTS

I don't know why people like the home run so much.
A home run is over as soon as it starts... wham,
bam, thank you, ma'am. The triple is the most
exciting play of the game. A triple is like meeting a
woman who excites you, spending the evening talking
and getting more excited, then taking her home. It
drags on and on. You're never sure how it's going
to turn out.

GEORGE FOSTER

There are so many wonderful theorems in this field from which to choose that important ones are sure to be passed over. Nevertheless, we have selected a few topics because they seem to be of a fundamental nature in that they reveal significant properties of random graphs. Furthermore, these topics have been actively investigated, and there are some nice results. As usual, the influence of the founding fathers, Erdös and Rényi, is apparent in studies of the degree distribution, the chromatic number and the clique number.

5.1. THE DEGREE DISTRIBUTION

There have been many contributors to this topic, including Erdös, Rényi, Ivčhenko and, most recently, Bollobás, who has made an intensive investigation of degree sequences. Some of the remarkable results are sketched below.

63

We have already found a sharp threshold for vertices of degree zero in Section 3.2. That result was completely generalized by Erdös and Rényi in their study of connectivity [ErR61]. For each nonnegative integer d, let $X_d(G)$ be the number of vertices in G of degree d. Then in Model A,

$$(5.1.1) \qquad E(X_d) = n\binom{n-1}{d} p^d (1-p)^{n-1-d}.$$

If d is fixed and $p^2 n \to 0$, we have

$$(5.1.2) \qquad E(X_d) \sim n(np)^d\left(\frac{e^{-pn}}{d!}\right).$$

Now just express p in the form $p = a_n (\log n)/n$ and set the right side of (5.1.2) equal to $e^{-x}/d!$. The result is

$$(5.1.3) \qquad a_n = 1 + d\frac{\log a_n}{\log n} + d\frac{\log\log n}{\log n} + \frac{x}{\log n},$$

and so

$$(5.1.4) \qquad p = \frac{\log n}{n} + d\frac{\log\log n}{n} + \frac{x}{n} + d\frac{\log a_n}{n}.$$

If we assume $a_n \to 1$, we obtain the threshold function in the next theorem [ErR61].

Theorem 5.1.1.   Let d be a fixed nonnegative integer. In Model A with the probability of an edge given by

$$(5.1.5) \qquad p = \frac{\log n}{n} + d\frac{\log\log n}{n} + \frac{x}{n} + o\left(\frac{1}{n}\right),$$

or in Model B with the number of edges given by

$$(5.1.6) \qquad q \sim \frac{1}{2}n\log n + \frac{d}{2}n\log\log n + \frac{x}{2}n + o(n),$$

the random variable X_d that counts vertices of degree d is distributed according to Poisson's law; that is, for each $k = 0, 1, 2, \ldots,$

$$P(X_d = k) \to e^{-\mu}\frac{\mu^k}{k!}$$

with mean $\mu = e^{-x}/d!$.

The proof is done in exactly the same manner as for Theorem 3.2.1, but the details are more involved (see exercise 5.1.3). In fact, X_d has a Poisson distribution whenever $\lim E(X_d)$ is finite, even if $d = d(n)$ as well [Bo82a]. For example [Pal-U], with constant $c > 0$ and $d > 0$, take

(5.1.7)
$$p = \frac{c}{n^{(d+1)/d}}.$$

Then $E(X_d) \sim c^d/d!$, and so the distribution is again Poisson in the limit. Palka [Pal84] also determined a threshold slightly below (5.1.5) for which the distribution is normal.

We use κ, λ an δ to denote the random variables for vertex connectivity, edge connectivity and minimum degree, respectively. Theorem 3.2.1 shows that $P(\delta > 0) \to \exp\{-\exp[-x]\}$, while Corollary 4.3.1 does the same for $P(\kappa > 0)$. Thus $P(\delta = 0)$ and $P(\kappa = 0)$ have the same limit, namely, $1 - \exp\{-\exp[-x]\}$. The next theorem [ErR61] generalizes this result.

Theorem 5.1.2. Let d be a fixed nonnegative integer. With the same thresholds as in the previous theorem, all three probabilities $P(\kappa = d)$, $P(\lambda = d)$ and $P(\delta = d)$ have the same limit, namely, $1 - e^{-\mu}$, where $\mu = e^{-x}/d!$.

The thresholds of Theorems 5.1.1 and 5.1.2 occur at a fairly advanced stage of the evolution. They are well beyond the arrival of the giant, and connectivity and minimum degree are gradually increasing for almost all graphs. To show that almost every graph has minimum degree $\geq d$, we just need $E(X_{d-1}) \to 0$. This expectation will tend to zero if we use p from (5.1.5) but replace d by $d - 1$ and x by a function ω_n that goes to infinity arbitrarily slowly:

(5.1.8)
$$p = \frac{\log n}{n}\left(1 + (d - 1)\frac{\log \log n}{\log n} + \frac{\omega_n}{\log n}\right).$$

Now with the probability of an edge given by (5.1.8), almost every graph has minimum degree and connectivity d.

We have already mentioned that the threshold that guarantees minimum degree 2 serves as a threshold for hamiltonicity [see formula (4.3.26)]. The minimum degree thresholds in Models A and B also are sufficient for almost all graphs to possess regular spanning subgraphs. For example, if the threshold of (5.1.8) is translated to Model B with $d = 1$, we have

(5.1.9)
$$q \sim \tfrac{1}{2}n \log n + \omega_n n,$$

which is exactly the threshold found by Erdös and Rényi [ErR66] for the existence of a 1-factor in a random graph. The proof of this result requires some fairly heavy going and relies on Tutte's characterization of graphs with 1-factors (see Appendix VIII).

Shamir and Upfal [ShU81] followed a rather different approach, using the augmentation of subfactors by alternating paths, a technique used in algorithmic studies of matching and flow problems [La76]. They found the following best possible theorem for spanning subgraphs with a specified degree sequence.

Theorem 5.1.3. Let $d \geq 1$ be a fixed positive integer. For each n, the probability of an edge is given by (5.1.8), and the degree sequence $d_1, \ldots, d_n$ of a graph is specified with $1 \leq d_i \leq d$ for $i = 1$ to n. Then in Model A almost every graph contains a subgraph with this specified degree sequence.

Hence if we take $d_i = d$ for $i = 1$ to n and consider only those values of n for which dn is even, then formula (5.1.8) is the threshold that ensures that almost every graph has a spanning subgraph that is regular of degree d.

More information about the distribution of the degrees when $pn/\log n$ is bounded away from 0 and ∞ can be found in the article of Bollobás [Bo82a]. But we will now turn to the more advanced phase in which $p = \omega_n(\log n)/n$ with $\omega_n \to \infty$ arbitrarily slowly. Erdös and Rényi [ErR60] found that in this case the degrees of all the vertices are very near the average degree. We use Δ to denote the maximum degree of a graph.

Theorem 5.1.4. Let $\varepsilon > 0$ be chosen and suppose $\omega_n \to \infty$ arbitrarily slowly. If the probability of an edge is

$$(5.1.10) \qquad p = \omega_n \frac{\log n}{n},$$

then

$$(5.1.11) \qquad P\big((1 - \varepsilon)pn < \delta \leq \Delta < (1 + \varepsilon)pn\big) \to 1,$$

that is, almost every graph satisfies the condition

$$(5.1.12) \qquad (1 - \varepsilon)pn < \deg v < (1 + \varepsilon)pn$$

for *each* of its vertices v.

Proof. Let $X(G)$ be the number of vertices of G whose degrees lie outside the interval of (5.1.12). Then the expectation of X is

$$(5.1.13) \qquad E(X) = n \sum \binom{n-1}{k} p^k (1 - p)^{n-1-k}$$

where the sum is over all k such that

$$(5.1.14) \qquad |k - pn| \geq \varepsilon pn.$$

Now we use the bounds on the upper and lower tails of the binomial distribution given in Proposition 4.1 of Appendix IV with $r = \lfloor (1 + \varepsilon) pn \rfloor$ and $s = \lceil (1 - \varepsilon) pn \rceil$. The result is

$$(5.1.15) \qquad E(X) = O(n)b(r; n - 1, p) + O(n)b(s; n - 1, p).$$

Since $pn \to \infty$ and $p^2 n \to 0$, we can use the bounds provided in formula (4.18) of Appendix IV for the right side of (5.1.15):

$$E(X) = O(n)(pn)^{-1/2}\exp\left\{-\frac{\varepsilon^2 pn}{3}\right\}$$

$$(5.1.16) \qquad = \frac{O(n)}{(\omega_n \log n)^{1/2} n^{\varepsilon^2 \omega_n /3}}$$

$$= o(1).$$

Since $E(x) \to 0$, almost every graph has no vertices outside the interval (5.1.10). □

Of course, stronger results could be stated. For example, (5.1.16) indicates that we could have let ε slowly approach zero and would still have reached the same conclusion (see Ivčhenko [Iv73], p. 189).

Ivčhenko studied in some detail the distribution of the mth largest and mth smallest degree. Here is his estimate of the interval $\Delta - \delta$, which generalizes the previous theorem.

Theorem 5.1.5. Let x be any real number and suppose $p = \omega_n (\log n)/n$ with $\omega_n \to \infty$ arbitrarily slowly. Then

$$(5.1.17) \qquad P\left(\left\{\Delta - \delta - 2\sqrt{\frac{2}{\omega_n}} [1 + o(1)]\right\} \sqrt{\frac{2}{\omega_n}} + \log 4\pi \le x\right)$$

$$= \int_0^\infty \exp\left\{-y - \frac{e^{-x}}{y}\right\} dy + o(1).$$

If p is fixed, more routine versions of the central limit theorem can be used to find bounds for the maximum and minimum degrees of almost all graphs. For example, Proposition 4.2 of the Appendix shows that the expected number of vertices of degree $k \ge \lfloor pn + x\sqrt{p(1 - p)n} \rfloor$ is $O(1)(ne^{-x^2/2})/x$. To make this last term go to zero, we just need $ne^{-x^2/2} = 1$ or $x = \sqrt{2 \log n}$.

Therefore almost every graph has

$$(5.1.18) \qquad \Delta < \left\lfloor pn + \sqrt{2p(1-p)n\log n} \right\rfloor.$$

Erdös and Wilson [ErW77] noticed that when $p = 1/2$, almost every graph has a unique vertex of maximum degree. This can be proved rather easily for any fixed p [Bo79]. The second-moment method is used to show that for any small $\varepsilon > 0$, almost every graph has a vertex of degree at least $\left\lfloor pn + (\sqrt{2} - \varepsilon)\sqrt{p(1-p)n\log n} \right\rfloor$. This means that

$$(5.1.19) \qquad \Delta = \left\lfloor pn + \sqrt{2p(1-p)n\log n} + o(n\log n)^{1/2} \right\rfloor.$$

Finally one shows that the expected number of pairs of vertices whose degrees are at least $\left\lfloor pn + c\sqrt{p(1-p)n\log n} \right\rfloor$ for $c > \sqrt{3/2}$ is $o(1)$. This is quickly accomplished by means of formula (4.24) of Appendix IV.

But the papers of Bollobás [Bo81b] and [Bo82a] go much further. For example, in [Bo82a] it is shown that for $p \leq 1/2$, almost every graph has a unique vertex of maximum (minimum) degree iff $pn/\log n \to \infty$. A detailed description of the degree sequence $d_1 \geq d_2 \geq \cdots \geq d_n$ of a random graph can be found in [Bo81b] for p fixed.

Sharper estimates are given for the lengths of the interval containing the mth highest degree of almost every graph. There is also a thorough treatment of the jumps $d_m - d_{m+1}$ and multiplicities of the degrees that occur in almost all graphs. Here is just one result of [Bo81b], which shows how many of the vertices of high degree are different besides the one of maximum degree.

Theorem 5.1.6.   In Model A with p fixed, if $m = o(n^{1/4})/(\log n)^{1/4}$, then almost every graph has $d_1 > d_2 > \cdots > d_m$, that is, the degrees of the m vertices of highest degree are all different, but if $m \neq o(n^{1/4})/(\log n)^{1/4}$ then almost every graph is such that $d_i = d_{i+1}$ for some $i < m$.

Earlier a slightly weaker form of this result was found by Babai, Erdös and Selkow [BaES80] together with a surprising application. There is no known polynomial–time algorithm for testing graphs for isomorphism [ReC77]. Nevertheless, Babai et al. found for almost all graphs that a very simple algorithm will test a given graph for isomorphism within linear [i.e., $O(n^2)$] time.

A *canonical labeling algorithm* creates a special class $\mathscr{H}$ of labeled graphs of order n. When applied to a graph G of order n, it either rejects G or gives the vertices new labels from 1 to n and puts G in the class $\mathscr{H}$. The new labeling is called *canonical*, and $\mathscr{H}$ is required to have two important properties:

(i) $\mathscr{H}$ is closed under isomorphism; that is, if two graphs are isomorphic and one is in $\mathscr{H}$, so is the other.

(ii) Two graphs in $\mathscr{H}$ are isomorphic iff they have the same canonical labeling; that is, the correspondence that preserves the canonical labels is an isomorphism.

Babai, Erdös and Selkow found the following beautiful, simple, effective algorithm. One can assume that the graph to which the algorithm is applied is represented by its adjacency matrix.

CANONICAL LABELING ALGORITHM

STEP 1. Compute $r = \lfloor 3 \log_2 n \rfloor$.

STEP 2. Compute the degree of each vertex of G.

STEP 3. Order the vertices by degree, calling them $v_1, v_2, \ldots, v_n$ so that we have

$$\deg v_1 \geq \cdots \geq \deg v_n.$$

STEP 4. If $\deg v_i = \deg v_{i+1}$ for any $i = 1$ to $r - 1$, reject G. Otherwise continue.

STEP 5. Rearrange rows and columns in the adjacency matrix of G so that $a_{i,j} = 1$ if v_i and v_j are adjacent and $a_{i,j} = 0$ otherwise. For $i = r + 1$ to n, compute

$$f(v_i) = \sum_{j=1}^{r} a_{i,j} 2^j,$$

called the "code of v_i with respect to $v_1, \ldots, v_r$."

STEP 6. Order the vertices $v_{r+1}, \ldots, v_n$ according to their f-value, calling them $w_{r+1}, \ldots, w_n$ so that

$$f(w_{r+1}) \geq \cdots \geq f(w_n).$$

STEP 7. If $f(w_i) = f(w_{i+1})$ for any $i = r + 1$ to $n - 1$, reject G. Otherwise continue.

STEP 8. Label vertex v_i with label i for $i = 1$ to r, and label vertex w_i with label i for $i = r + 1$ to n. This labeling is called canonical.

Thus the r vertices of highest degree are arranged in descending order of degree and given the labels 1 to r. Then the other $n - r$ vertices are arranged in an order that depends on the code determined by their adjacencies with the vertices of high degree and are assigned the remaining labels. If there are any ties during this process, that is, if two vertices deserve the same label, then the graph is rejected.

It is easy to see that $\mathscr{H}$ satisfies the two properties of a canonical labeling algorithm. Furthermore, the complexity of the algorithm is $O(n^2)$, since the most costly Step 5 may require the rearrangement of about n rows and columns of the adjacency matrix.

Here is the algorithm for testing two given labeled graphs G_1 and G_2 of order n for isomorphism.

ISOMORPHISM ALGORITHM

STEP 1.   Apply the canonical labeling algorithm to G_1 and G_2. Quit if the algorithm rejects either graph.

STEP 2.   Consider the correspondence between the vertices of G_1 and G_2 determined by the canonical labelings. The two graphs are isomorphic iff this correspondence determines an isomorphism, so this step could be completed by comparing the two adjacency matrices whose rows and columns are indexed by the canonical labelings. The two graphs are isomorphic iff these two matrices are identical.

The main result of [BaES80] showed that this algorithm almost always works! For this it is sufficient to show that the canonical labeling algorithm hardly ever rejects a graph.

Theorem 5.1.7.   In Model A with $p = 1/2$, the probability $P(\mathscr{H})$ of the set of canonically labeled graphs of order n is greater than $1 - \sqrt[7]{1/n}$ for sufficiently large n, that is,

(5.1.20) $P(\mathscr{H}) \to 1,$

and almost all graphs have a canonical labeling.

Note that Theorem 5.1.6 shows that the degrees of the $\lfloor 3 \log_2 n \rfloor$ vertices of highest degree are all different for almost all graphs. Therefore almost all graphs will reach Step 5 of the canonical labeling algorithm. The tough part of the proof shows that the remaining vertices can almost always be distinguished by their adjacencies to the vertices of high degree.

The authors stress that their algorithm is not intended for practical use, but just to show that a simple algorithm almost always works. There are other algorithms that are more complicated but also more successful. Babai et al. also suggested some problems that lead to the following interesting results. R. Lipton [Li-U] found a canonical labeling algorithm of complexity $O(n^6 \log n)$ for which the probability of rejection is exponentially small ($c^{-n}, c > 1$). This was improved by R. M. Karp [Ka79] with an $O(n^2 \log n)$ algorithm that has probability of rejection $O(n^{3/2}2^{-n/2})$. Babai and Kučera [BaK79] showed that the labeling algorithm just given has an exponentially small probability of rejection and also proved that the rejected graphs can be salvaged so that the canonical labeling algorithm works on *all* labeled graphs of order n with expected time $O(n^2)$!

We conclude with another application, which involves the edge chromatic number of a graph. Vizing has shown (see Appendix VIII for graph theory) that the edge chromatic number of any graph is either the maximum degree Δ or $\Delta + 1$. He also found that every graph that requires $\Delta + 1$ colors has at least three vertices of maximum degree. As mentioned earlier, Erdös and Wilson found that almost all graphs have a unique vertex of maximum degree [see formula (5.1.19)]. Thus they could conclude that almost all graphs have edge chromatic number Δ.

Theorem 5.1.8. In Model A with p fixed, almost all graphs have edge chromatic number equal to the maximum degree.

But let us now consider only those graphs of order n with *fixed* maximum degree m. As n tends to infinity, what is the limiting probability that these graphs have edge chromatic number m? A graph of maximum degree 2 consists of components that are paths and cycles. If there is one odd cycle, the graph has edge chromatic number 3. Exponential generating functions (see [HP73], Chapter 1) can be used to derive recurrence relations for the number of labeled graphs of maximum degree 2 as well as for those with no odd cycles. For $n = 3$ to 10 vertices, the probability of no odd cycles is between .8 and .9 and it seems likely to converge to about .85. For $m \geq 3$, however, the problem of determining the proportion of graphs with edge chromatic number m remains open.

Exercises 5.1

1. (a) If $X(G) = \deg v_1$, the degree of the first vertex, find $E(X)$ in Model B.

   (b) How does $P(X = k)$ behave asymptotically in Model B for k fixed and $q = o(n^{3/2})$?

2. What can be concluded about the degree of a given vertex using Chebyshev's inequality?

3. Work out $E(X_d^2)$ in Model A just for $d = 1$. When the probability of an edge is $p = 2c/n$ (the land of the giant for $2c > 1$), show that there are certain to be vertices of degree 1.

4. What is the inequality corresponding to (5.1.18) for the minimum degree δ?

5. Let $X_m(G)$ be the number of m-sets of vertices whose deletion disconnects G. In Model A with p fixed and $m = m(n)$, show that
    (a) $E(X_m) = O(1)\binom{n}{m}(1 - p)^{n-m}$.
    (b) $E(X_m) \to 0$ for $m = (1 - \varepsilon)pn$ provided $\varepsilon > 0$ is sufficiently close to 1.

Hence almost all graphs are m-connected.

5.2. THE CHROMATIC NUMBER

Not much is known about the chromatic number of a random graph except in the early period of the evolution when the giant has just arrived and in the twilight zone when the probability of an edge is fixed. From Section 4.1 we know that in Model B with $q = o(n)$ the components of almost every graph are trees. Therefore these graphs all have chromatic number 2.

If $q \sim cn$ with $0 < c < 1/2$, Erdös and Rényi (Theorem 4.2.6) showed that all components are trees or contain exactly one cycle. For such a graph the chromatic number will be 2 unless there is an odd cycle, in which case it will be 3. Hence, $\chi \leq 3$.

If $q \sim \frac{1}{2}n$, Erdös and Rényi observe that one can show that almost every graph contains an odd cycle and the proof is done in the same way as for Theorem 4.1.1. Therefore $\chi \geq 3$ for almost all of these graphs.

In Model A with $p = 2c/n$, Erdös and Spencer [ErS74] have stated that for large c almost all graphs G have

$$(5.2.1) \qquad \frac{c}{\log c}[1 + o(1)] \leq \chi(G) \leq \frac{2c}{\log c}[1 + o(1)]$$

where $o(1)$ is with respect to c. They conjecture that the same is true if $c = c(n) \to \infty$, $c(n) = o(n)$ and $o(1)$ is with respect to n.

The most interesting results on the chromatic number occur in Model A with p fixed. Grimmett and McDiarmid [GrM75] found the bounds in the next theorem (see also [ErS74], p. 56), and these results were subsequently sharpened in [BoE76].

The lower bound will be derived from the inequality for graphs of order n:

$$(5.2.2) \qquad \chi \geq \frac{n}{\beta},$$

where β is the independence number, that is, the maximum number of mutually nonadjacent (independent) vertices.

The upper bound is established by estimating the effectiveness of the greedy algorithm, sometimes called the sequential coloring algorithm (see [MaMI72] for an interesting discussion and empirical evidence). A list $c_1, c_2, \ldots$ of colors is available, and we consider any labeled graph G of order n. We always assume that the vertex with label i is denoted by v_i for $i = 1$ to n. The algorithm first colors vertex v_1 with color c_1. Now suppose vertices $v_1, \ldots, v_m$ have been colored properly, that is, adjacent vertices have been assigned different colors. Then vertex v_{m+1} is given the available color with the lowest possible subscript. For example, if v_2 is not adjacent to v_1, then v_2 gets color c_1; otherwise v_2 gets color c_2, and so on. Thus the number of colors used is completely determined by the labels and is an upper bound for χ.

Theorem 5.2.1. Let $0 < \varepsilon < 1$ be given, and set $b = 1/(1 - p)$. In Model A with p, the probability of an edge, fixed, the chromatic number of almost all graphs G satisfies

$$(5.2.3) \qquad \frac{\left(\frac{1}{2} - \varepsilon\right)n}{\log_b n} \le \chi(G) \le \frac{(1 + \varepsilon)n}{\log_b n}.$$

Proof. First we will deal with the lower bound. We need a function $r = r(n)$ such that $\beta \le r(n)$ for almost all graphs. Then (5.2.2) implies that $\chi \ge n/r(n)$.

Therefore we define the random variable $X_r(G)$ to be the number of r-sets of independent vertices. Then the expected value is

$$(5.2.4) \qquad E(X_r) = \binom{n}{r}(1 - p)^{\binom{r}{2}} \le \left[n(1 - p)^{(r-1)/2}\right]^r.$$

To make $E(X_r) \to 0$, we just need $n(1 - p)^{(r-1)/2} \le 1 - \delta$. Taking logs and solving for r, we find for any $\varepsilon_1 > 0$ that

$$(5.2.5) \qquad r = 2(1 + \varepsilon_1)\log_b n$$

will do the job. Since almost every graph has no independent r-sets for this r, we must also have $\beta < 2(1 + \varepsilon_1)\log_b n$. Hence

$$\chi \ge \frac{1}{2(1 + \varepsilon_1)} \frac{n}{\log_b n},$$

which implies the lower bound in (5.2.3) on proper choice of ε_1.

Next we establish the upper bound (see [BoE76]). We wish to find a function $f(n)$ such that almost all graphs G have $\chi(G) \leq f(n)$. It would be sufficient to show that the greedy algorithm almost always uses $\leq f(n)$ colors. Therefore let $\mathscr{A}$ be the labeled graphs of order n for which the greedy algorithm uses $\leq f(n)$ colors. These are the "good" graphs. To show $P(\mathscr{A}) \to 1$, we will, of course, want to prove that $P(\overline{\mathscr{A}}) \to 0$.

Let G be a labeled graph of order n. Suppose the greedy algorithm uses $\leq f(n)$ colors on the first m vertices $v_1, \ldots, v_m$ but requires $> f(n)$ colors when v_{m+1} is included. That is, v_{m+1} is assigned color c_j, $j = f(n) + 1$. Then v_{m+1} is a "bad" vertex, G is a "bad" graph and

$$(5.2.6) \qquad P(\overline{\mathscr{A}}) = \sum_{m=1}^{n-1} P(\mathscr{B}_{m+1}),$$

where $\mathscr{B}_{m+1}$ is the set of graphs of order n for which vertex v_{m+1} is bad.

Now we need an upper bound for $P(\mathscr{B}_{m+1})$. For any graph G of order n, we denote the subgraph induced by the set $\{v_1, \ldots, v_m\}$ of vertices by $G[v_1, \ldots, v_m]$. Suppose H is a graph of order m with vertex set $v_1, \ldots, v_m$ and that the greedy algorithm, when applied to H, uses exactly $f(n)$ colors. Therefore for $i = 1$ to $f(n)$, these are $k_i \geq 1$ vertices that are assigned color c_i and $\sum_{i=1}^{f(n)} k_i = m$. Then the probability of all graphs G in $\mathscr{B}_{m+1}$ such that $G[v_1, \ldots, v_m]$ is isomorphic to H is

$$(5.2.7) \qquad P(H) \prod_{i=1}^{f(n)} \left[1 - (1-p)^{k_i} \right],$$

because v_{m+1} must be adjacent to at least one vertex of each color in H. But a neat inequality of [BoE76] (see exercise 5.2.2) shows that

$$(5.2.8) \qquad \prod_{i=1}^{f(n)} \left[1 - (1-p)^{k_i} \right] \leq \left[1 - (1-p)^{n/f(n)} \right]^{f(n)}.$$

Now that we have a bound on the product that is independent of the k_i, we can sum (5.2.7) over *all* graphs H of order m and obtain the following bound on the probability that the greedy algorithm needs $> f(n)$ colors for v_{m+1}:

$$(5.2.9) \qquad P(\mathscr{B}_{m+1}) \leq \left[1 - (1-p)^{n/f(n)} \right]^{f(n)}.$$

Of course this bound is also independent of m, and so from (5.2.6) it follows that

$$(5.2.10) \qquad \begin{aligned} P(\overline{\mathscr{A}}) &\leq n \left[1 - (1-p)^{n/f(n)} \right]^{f(n)} \\ &\leq n \exp\left\{ -f(n)(1-p)^{n/f(n)} \right\}. \end{aligned}$$

It remains to choose $f(n)$ so that the right side of (5.2.10) goes to zero and, of course, we want $f(n)$ as low as possible. Since we already have a lower bound, it is reasonable to try an upper bound of the same form. So we let $f(n) = cn/\log_b n$, substitute in the right side of (5.2.10) and take logs. The result is

(5.2.11) $$\log P(\bar{\mathscr{A}}) \le \log n - \frac{cn^{1-1/c}}{\log_b n}.$$

The right side of (5.2.11) goes to $-\infty$ iff $1 - 1/c > 0$, and this leads to the choice of $f(n) = (1 + \varepsilon)n/\log_b n$ in the statement of the theorem.   □

The bounds in the theorem for χ are so "close" that it may be tempting to try to tighten the gap. However, lowering the upper bound on χ may require a substantial improvement in the greedy algorithm. There is no known efficient algorithm for determining χ in general; in fact, the graph coloring problem is NP-complete. Making even a slight improvement of the upper bound may be as hard as proving that $P = \text{NP}$! There is some evidence to support this view [GaJ76]. Suppose we were to find a nice improvement on the greedy algorithm that used $\le (1 - \varepsilon)n/\log_b n$ colors for almost all graphs. But multiplying the lower bound of Theorem 5.2.1 by 2 gives $(1 - \varepsilon)n/\log_b n \le 2\chi$, and this in turn means that the improved algorithm uses fewer than 2χ colors. However, the main result of [GaJ76] indicates that there is then a polynomial time algorithm for determining χ!

Exercises 5.2

1.   Construct a labeled graph of order $2n$ and chromatic number 2 for which the greedy algorithm uses n colors.

2.   If $0 < \alpha < 1$ and the positive integers k_i for $i = 1$ to $f(n)$ satisfy $\sum_{i=1}^{f(n)} k_i \le n - 1$, then

$$\prod_{i=1}^{f(n)} (1 - \alpha^{k_i}) < (1 - \alpha^{n/f(n)})^{f(n)}.$$

5.3.   THE CLIQUE NUMBER

A *clique* of a graph is a maximal complete subgraph. The *clique number* of a graph G, denoted $\text{cl}(G)$, is the maximum order of a complete subgraph. One of the most surprising results in the theory of random graphs was discovered by

Matula [Ma72], who found that in Model A with p fixed the clique number of almost all graphs assumed one of just two possible values! Earlier [Ma70] he found the following empirical evidence that suggested this. With $p = 1/2$, 165 random graphs of order 32 were generated. Of these, 1 had clique number 5, 90 had 6, 68 had 7 and the remaining 6 had 8. Thus 158 out of 165, or over 95%, had clique numbers 6 or 7.

We know from Theorem 2.3.1 that if r is any fixed positive integer, then in Model A with p fixed almost every graph contains a complete subgraph of order r. To learn about the clique number, we need to know how fast r can be allowed to increase so that there is still almost always a complete subgraph of order r. Therefore we must consider the expected number $E(X_r)$ of complete subgraphs of order r:

$$(5.3.1) \qquad E(X_r) = \binom{n}{r} p^{\binom{r}{2}}.$$

We seek a slowly increasing function $r = r(n)$ such that $E(X_r) \to 0$. Using Stirling's formula, we find

$$(5.3.2) \qquad E(X_r) \sim \frac{1}{\sqrt{2\pi r}} \left(\frac{enp^{(r-1)/2}}{r} \right)^r.$$

Now observe that $E(X_r) \to 0$ if $r \to \infty$, and for large n

$$(5.3.3) \qquad \frac{enp^{(r-1)/2}}{r} \leq 1.$$

Taking the log of both sides of (5.3.3) we have

$$(5.3.4) \qquad r \geq 2\frac{\log n}{\log b} - 2\frac{\log r}{\log b} + 1 + \frac{2}{\log b},$$

where $b = 1/p$. Since the lower bound is asymptotic to $2\log_b n$, we substitute this value for r on the right side of (5.3.4) and find that we require (see Matula [Ma72])

$$(5.3.5) \qquad r \geq 2\log_b n - 2\log_b \log_b n + 1 + 2\log_b e - 2\log_b 2.$$

The right side of this inequality turns out to be the critical value for determining the clique number of almost every graph, as seen in the following statement of Matula's theorem.

Theorem 5.3.1. In Model A, the probability of an edge is fixed at $p = 1/b$, and $\varepsilon > 0$ is given. Let the function $d = d(n)$ be defined by

$$(5.3.6) \qquad d(n) = 2\log_b n - 2\log_b \log_b n + 1 + 2\log_b\left(\frac{e}{2}\right).$$

Then

(5.3.7) $$P(\lfloor d - \varepsilon \rfloor \leq \mathrm{cl}(G) \leq \lfloor d + \varepsilon \rfloor) \to 1,$$

that is, the clique number of almost every graph is between $\lfloor d - \varepsilon \rfloor$ and $\lfloor d + \varepsilon \rfloor$.

Proof. The first task is to show that $\mathrm{cl}(G) \leq \lfloor d + \varepsilon \rfloor$. It is sufficient to show that there are no complete subgraphs of order $> \lfloor d + \varepsilon \rfloor$; that is, $E(X_r) \to 0$ for $r \geq d + \varepsilon$. From what we did earlier, it follows that this part of the proof is completed by checking that $r = d + \varepsilon$ satisfies equation (5.3.4) for large n (see exercise 5.3.4).

The second task is to show that $\mathrm{cl}(G) \geq \lfloor d - \varepsilon \rfloor$. The second-moment method can be used to show that there is certain to be a complete subgraph of order $\geq \lfloor d - \varepsilon \rfloor$. Here is a sketch of the procedure. First we need the expectation of X_r^2:

(5.3.8) $$E(X_r^2) = \sum_{k=0}^{r} \binom{n}{r}\binom{r}{k}\binom{n-r}{r-k} p^{2\binom{r}{2}-\binom{k}{2}}.$$

From (5.3.1) and (5.3.8) we have the ratio

(5.3.9) $$\frac{E(X_r^2)}{E(X_r)^2} = a_n + b_n$$

where

(5.3.10) $$a_n = \binom{n}{r}^{-1}\left[\binom{n-r}{r} + \binom{r}{1}\binom{n-r}{r-1}\right]$$

and

(5.3.11) $$b_n = \binom{n}{r}^{-1}\sum_{k=2}^{r} \binom{r}{k}\binom{n-r}{r-k} b^{\binom{k}{2}}.$$

Formula (3.8) of Appendix III can be used to show that $a_n \sim 1$ for $r \sim 2\log_b n$. To complete the proof, it must be shown that $b_n \to 0$ when $r = \lfloor d - \varepsilon \rfloor$. Consider, for example, the top term ($k = r$) in the sum for b_n. It is exactly $1/E(X_r)$, and we can use (5.3.2) to determine its behavior:

(5.3.12) $$\frac{1}{E(X_r)} \sim \sqrt{2\pi r}\left(\frac{rb^{(r-1)/2}}{en}\right)^r.$$

Using the hypothesis that $r = \lfloor d - \varepsilon \rfloor \le d - \varepsilon$, we find

$$(5.3.13) \qquad \frac{rb^{(r-1)/2}}{en} \le \frac{r}{2 \log_b n} b^{-\varepsilon/2} \sim b^{-\varepsilon/2}.$$

And so $1/E(X_r) \to 0$ neatly for small $\varepsilon > 0$.

The behavior of the top term gives the clue to the behavior of the top part of the sum in b_n. To see this, let b_n^+ denote the sum in b_n from $k = m$ to r:

$$b_n^+ = \binom{n}{r}^{-1} \sum_{k=m}^{r} \binom{r}{k}\binom{n-r}{r-k} b^{\binom{k}{2}}$$

$$(5.3.14)$$

$$= \binom{n}{r}^{-1} b^{\binom{r}{2}} \sum_{k=m}^{r} \binom{r}{k}\binom{n-r}{r-k} b^{\binom{k}{2}-\binom{r}{2}}$$

From (5.3.12) and (5.3.13) it follows that

$$(5.3.15) \qquad b_n^+ = O(1)\sqrt{r}\,(b^{-\varepsilon/2})^r F(m),$$

where

$$(5.3.16) \qquad F(m) = \sum_{k=m}^{r} \binom{r}{r-k}\binom{n-r}{r-k} b^{\binom{k}{2}-\binom{r}{2}}.$$

Now set $j = r - k$ and interchange the order of summation:

$$(5.3.17) \qquad F(m) = \sum_{j=0}^{r-m} \binom{r}{j}\binom{n-r}{j} b^{-\binom{j}{2}-j(r-j)}.$$

Using crude bounds for the binomial coefficients in (5.3.17), we have

$$(5.3.18) \qquad \binom{r}{j}\binom{n-r}{j} \le (rn)^j,$$

and hence

$$(5.3.19) \qquad F(m) \le \sum_{j=0}^{r-m} \left(\frac{rn}{b^{r-(j+1)/2}} \right)^j.$$

It can be seen that $F(m)$ will be bounded if j is not too close to r. Therefore we define m by choosing any α, $0 < \alpha < 1$, and setting $m = \lceil (1 - \alpha)r \rceil$. Then $j \le r - m \le \alpha r$, and $r - (j + 1)/2 \ge r(1 - \alpha/2) - 1/2$. These bounds can be applied to equation (5.3.19) to obtain

$$(5.3.20) \qquad F(m) \le \sum_{j=0}^{\infty} \left(\frac{rn\sqrt{b}}{b^{r(1-\alpha/2)}} \right)^{j},$$

and so to show that $F(m)$ is bounded we just need

$$(5.3.21) \qquad \frac{rn\sqrt{b}}{b^{r(1-\alpha/2)}} < 1$$

for large n. To do this, recall that $r = \lfloor d - \varepsilon \rfloor \ge (d - \varepsilon) - 1$. Therefore $b^r \ge b^{d-\varepsilon-1}$. Using (5.3.6), where d is defined, we have the following estimate for the left side of (5.3.21):

$$
\begin{aligned}
(5.3.22) \qquad \frac{rn\sqrt{b}}{b^{r(1-\alpha/2)}} &\le \frac{rn}{(n/\log_b n)^{2-\alpha}} \frac{\sqrt{b}}{\left[(e/2)^2 b^{-\varepsilon} \right]^{1-\alpha/2}} \\
&= O(1) \frac{(\log_b n)^{3-\alpha}}{n^{1-\alpha}} \\
&= o(1).
\end{aligned}
$$

Therefore for any α, we have $F(m) = O(1)$, and so $b_n^+ \to 0$.

Now we turn to the bottom part of the sum in b_n and let

$$(5.3.23) \qquad b_n^- = \binom{n}{r}^{-1} \sum_{k=2}^{m} \binom{r}{k} \binom{n-r}{r-k} b^{\binom{k}{2}}.$$

As before it is helpful to notice the behavior of the bottom term in this sum. Then we know that we should first observe that

$$
\begin{aligned}
(5.3.24) \qquad \binom{n}{r}^{-1} \binom{r}{k} \binom{n-r}{r-k} &= \frac{r!}{(n)_r} \frac{(r)_k}{k!} \frac{(n-r)_{r-k}}{(r-k)!} \\
&= \frac{r!}{(r-k)!} \frac{(n-r)_{r-k}}{(n)_r} \frac{(r)_k}{k!} \\
&\le r^k \frac{1}{(n)_k} \frac{r^k}{k!} \sim \frac{r^{2k}}{k! n^k},
\end{aligned}
$$

where the last step follows from formula (3.5) of the Appendix since $k \le r = o(\sqrt{n})$. Returning to (5.3.23) we can write

$$(5.3.25) \qquad b_n^- = O(1) \sum_{k=2}^{m} \left(\frac{r^2 b^{k/2}}{n\sqrt{b}} \right)^{k},$$

and we see that $b_n^- \to 0$ provided $b^{k/2}$ is not too large. Therefore we set $m = \lfloor \beta r \rfloor$ with $0 < \beta < 1$ and try to show that

(5.3.26) $$\frac{r^2 b^{k/2}}{n\sqrt{b}} = o(1)$$

for $2 \le k \le \lfloor \beta r \rfloor$.

Since $k \le \beta r \le \beta(d - \varepsilon)$,

$$\frac{r^2 b^{k/2}}{n\sqrt{b}} \le \frac{r^2}{n}\left(\frac{n}{\log_b n}\right)^\beta \frac{\left(\sqrt{b}\,(e/2)b^{-\varepsilon/2}\right)^\beta}{\sqrt{b}}$$

(5.3.27)
$$= O(1)\frac{(\log_b n)^{2-\beta}}{n^{1-\beta}}$$

$$= o(1).$$

Since both α and β were arbitrary, we simply select them such that our estimates of the upper and lower portions of the sums in b_n overlap. We just need $\lfloor \beta r \rfloor \ge \lceil (1 - \alpha)r \rceil$ or $\alpha + \beta \ge 1$ to ensure that $b_n \to 0$.   □

This result was sharpened in [BoE76], where one can find other results about the distribution of the clique number.

Other types of clique numbers have also been studied, such as the topological clique number [ErF81] and the contraction clique number [BoCE80]. This latter invariant, denoted ccl(G), is the largest integer r such that G can be contracted to a complete graph of order r, that is, a sequence of edge shrinkings reduces G to K_r. The bounds on the contraction clique number found by [BoCE80] for almost all graphs are given by the following theorem.

Theorem 5.3.2. In Model A with p fixed and $b = 1/(1 - p)$, the contraction clique number satisfies the following inequalities for almost all graphs:

(5.3.28) $\quad n\left[(\log_b n)^{1/2} + 1\right]^{-1} \le \text{ccl}(G) \le n\left[(\log_b n)^{1/2} - 1\right]^{-1}.$

Bollobás, Catlin and Erdös used the lower bound above to derive the following exciting corollary. The long-standing conjecture of Hadwiger states that if a graph G has chromatic number r, then it can be contracted to a complete graph of order r; that is,

(5.3.29) $$\chi(G) \le \text{ccl}(G).$$

Corollary 5.3.1. Hadwiger's conjecture is true for almost every graph!

From the lower bound in (5.3.28) and the upper bound in (5.2.3) for $\chi(G)$, we have for any $\varepsilon > 0$,

$$(5.3.30) \qquad \frac{\chi(G)}{\mathrm{ccl}(G)} \leq (1 + \varepsilon)\frac{1 + (\log_b n)^{1/2}}{\log_b n} = o(1)$$

for almost all graphs. Hence

$$(5.3.31) \qquad P\left(\frac{\chi}{\mathrm{ccl}} \leq 1\right) \to 1,$$

and the corollary is verified.

Exercises 5.3

Let $X(G)$ be the number of cliques of order r in G.

1. Find $E(X)$ in Model A.

2. Find $E(X)$ in Model B.

3. Find a threshold in Model A for the property that the clique number is at least 4.

4. Verify that $r = d + \varepsilon$ satisfies equation (5.3.4) for large n.

5. With the same hypothesis as exercise 2.3.5, express k as an explicit function of p, n and the new variable x so that the bad k-sets are distributed in the limit according to Poisson's law.

5.4. STRENGTH OF A THEOREM

The effectiveness of a theorem about graphs of order n and size q could be measured by enumeration. Just divide the number of graphs that satisfy the hypothesis by the number that satisfy the conclusion. The closer the resulting quotient is to 1, the wider is the applicability of the theorem and the closer the hypothesis is to a characterization. On the other hand, if the measure is close to zero, the hypothesis rarely holds. Note that some perfectly respectable propositions may fall into this category. For example, consider the following simple but useful fact.

Proposition 1. Every tree has at least two vertices of degree 1.

This is a proposition about graphs of order n and size $q = n - 1$. The number of these that satisfy the hypothesis is provided by Cayley's theorem, that is, there are n^{n-2} labeled trees of order n. For an estimate of the number of graphs of order n and size $n - 1$ with vertices of degree 1, we can apply the theory of evolution. We are considering size $q = n - 1 \sim cn$ with $c = 1 > 1/2$. Therefore we are in the phase of growth for the giant component, and Theorem 4.2.4 shows the proportion of vertices outside the giant that belong to trees. In fact, exercise 5.1.3 explicitly demonstrates the existence of vertices of degree 1 for any $c > 0$ in Model A. So almost all graphs of order n and size $q = n - 1$ have vertices of degree 1. Exercise 5.4.1 requests a nice asymptotic estimate of these from which it easily follows that

(5.4.1)
$$\frac{n^{n-2}}{\left(\begin{array}{c}\binom{n}{2}\\n-1\end{array}\right)} \to 0,$$

that is, almost none of these graphs are trees. The proposition is not improved by including in the hypothesis all acyclic graphs. The number of these (see p. 29 of [M70]) is asymptotic to $n^{n-2}\sqrt{e}$, and once again almost no graph satisfies the hypothesis.

However, there is a very simple modification of the hypothesis suggested by Theorem 4.2.4 that strengthens the proposition about as much as possible.

Proposition 2. A graph of order n and size $q = n - 1$ with a component that is a tree has at least two vertices of degree 1.

The theory of graphical evolution tells us that if $q \sim cn$, almost all graphs have components that are trees. In fact they have isolated edges (see exercise 5.4.2).

Thus almost all graphs of order n and size $q = n - 1$ satisfy both the hypothesis and the conclusion of Proposition 2. Therefore Proposition 2 could be considered stronger than Proposition 1. On the other hand, there is much to say in favor of Proposition 1. In proving theorems about trees, one wants to know that they have vertices of degree 1 so that these can be deleted and an induction hypothesis applied. One does not care about random graphs of order n and size $q = n - 1$ for these purposes. Nevertheless, it may often be revealing to measure the strength of a theorem in this way, and so we shall discuss a few examples that seem particularly suited to this type of analysis.

The sharp threshold of Theorem 4.3.3 found by Komlós and Szemerédi [KoS83] places us in a favorable position for the analysis of sufficient conditions for a graph to be hamiltonian. The search for such conditions has produced enough interesting results over the last 40 years to provide ample material for discussion (see Appendix VIII).

One of many graph theory results that requires no induced copies of $K_{1,3}$ as subgraphs is the condition of Oberly and Sumner [ObS79]:

SC1. G has order $n \geq 3$, is connected and locally connected, and $K_{1,3}$ does not occur as an induced subgraph.

We have seen in Section 2.3 that in Model A with p fixed almost all graphs are connected and locally connected. However, almost all contain any given graph as an induced subgraph, and hence $K_{1,3}$ is almost always present. Thus almost no graph satisfies SC1. On the other hand, when the probability of an edge is fixed, it is well beyond the threshold (4.3.27) of Theorem 4.3.3, and so almost all graphs are hamiltonian.

But SC1 can still be useful. We can ask for its threshold in Model A, that is, for what $p = p(n) \to 0$ does SC1 almost always hold? In [ErPR83] we showed that if $p = [c(\log n)/n]^{1/2}$ with $c > 3/2$, then almost every graph is connected and locally connected. This function is a threshold for local connectivity, because if $0 < c \leq 3/2$, almost every graph is connected but has an edge that is not in a triangle (see exercise 3.1.5) and hence is not locally connected. But the threshold forbidding the little tree $K_{1,3}$ is much higher. Let $X(G)$ be the number of induced subgraphs of G that are isomorphic to $K_{1,3}$. Then in Model A

(5.4.2)
$$E(X) = \binom{n}{1, 3, n-4} p^3 (1-p)^3.$$

Now $E(X) \to 0$ if $n^4 p^3 \to 0$, but then almost no graph has components of order ≥ 4 (see exercise 4.1.4). So we must have $n^4(1-p)^3 \to 0$. This just means that $1 - p \to 0$ so fast that the complement of almost every graph has no components of order ≥ 4. Hence we need $p \to 1$ for certain application of SC1. Nevertheless there are a surprising number of graphs to which the theorem can be applied. I found that of the 156 unlabeled graphs of order 6, 15 of these satisfy SC1 while 48 are hamiltonian.

Next we consider the sufficient condition found by Dirac [Di52]:

SC2. G has order $n \geq 3$ and minimum degree $\delta \geq n/2$.

We want to find the probability of an edge such that SC2 holds for almost all graphs. Applying Theorem 5.1.4 with p defined by (5.1.10), we see that if

$$(5.4.3) \qquad\qquad (1 - \varepsilon) pn \geq \frac{n}{2},$$

then $\delta \geq n/2$. Therefore if $p > 1/2$ is fixed, ε can be chosen such that (5.4.3) holds. If $p < 1/2$, ε can be picked to make

$$(5.4.4) \qquad\qquad (1 + \varepsilon) pn < \frac{n}{2},$$

so that the maximum degree Δ is bounded above by $n/2$. Then almost no graph satisfies SC2.

When p is fixed, we have seen in Section 5.1 that almost every graph has a unique vertex of maximum (and minimum) degree. In fact,

$$(5.4.5) \qquad \delta = \left\lceil pn - [2p(1 - p)n \log n]^{1/2} + o(n \log n)^{1/2} \right\rceil,$$

and so if $p = 1/2$, there is a unique vertex of minimum degree

$$(5.4.6) \qquad\qquad \left\lceil n/2 - [(n \log n)/2]^{1/2} + o(n \log n)^{1/2} \right\rceil$$

for almost all graphs. It is easy to check that (5.4.6) is strictly less than $n/2$ for large n, and so SC2 holds for almost no graphs when $p \leq 1/2$.

Ore's condition [Or60] can be handled in much the same way as Dirac's.

SC3. G has order $n \geq 3$, and for every pair of nonadjacent vertices u and v,

$$(5.4.7) \qquad\qquad \deg u + \deg v \geq n.$$

If $p > 1/2$ is fixed, condition (5.4.7) will hold for almost all graphs because $\delta \geq n/2$. And if $p < 1/2$, almost all graphs have $\Delta < n/2$, and so (5.4.7) is almost never satisfied. To deal with $p = 1/2$ we define $X(G)$ to be the number of pairs of nonadjacent vertices in G that do not satisfy (5.4.7). Then the expectation of these bad pairs is

$$(5.4.8) \quad E(X) = \binom{n}{2}(1 - p) \sum_{k=0}^{n-1} \binom{2(n - 2)}{k} p^k (1 - p)^{2(n-2)-k}.$$

It is easy to verify (see exercise 5.4.3) that for $p = 1/2$

$$(5.4.9) \qquad\qquad E(X) \sim \frac{n^2}{8}.$$

Furthermore, the second-moment method shows that for $p = 1/2$ almost every graph has a bad pair of vertices that fail to satisfy (5.4.7). The details have been left for the reader to clean up in exercise 5.4.4. And so from a probabilistic point of view, Ore's condition SC3 is not significantly different from Dirac's SC2. And neither is Pósa's SC4 despite the fact that Pósa's theorem [Po62] is a substantial generalization of the others.

SC4. G is a graph of order $n \geq 3$ in which the degrees of the vertices listed in ascending order are $d_1 \leq d_2 \leq \cdots \leq d_n$, and for any integer k with $1 \leq k < (n - 1)/2$,

$$(5.4.10) \qquad k < d_k,$$

while if n is odd, for $k = (n + 1)/2$,

$$(5.4.11) \qquad k \leq d_k.$$

These two inequalities imply that about half the vertices of G have degree at least $n/2$. So once again, SC4 cannot hold for almost all graphs unless $p \geq 1/2$. And the same must be said for the more advanced degree requirements found by Bondy [Bo69], Chvátal [Ch72] and Las Vergnas [LV76]. However, there is a winner in these hamiltonian sweepstakes. It was discovered by Chvátal and Erdös [ChE72], and its beauty and simplicity are striking.

SC5. The connectivity of a graph G of order $n \geq 3$ is at least as large as the independence number, that is, $\kappa \geq \beta$.

It will be a cinch to show that in Model A, for *any* fixed $p > 0$, SC5 holds for almost all graphs because we already have the appropriate bounds on κ and β. From Theorem 2.3.2 we have for almost all graphs

$$(5.4.12) \qquad \kappa \geq o\left(\frac{n}{\log n}\right),$$

and while proving Theorem 5.2.1 we showed that for any $\varepsilon > 0$

$$(5.4.13) \qquad \beta < 2(1 + \varepsilon)\log_b n$$

for almost all graphs. Recall that $b = 1/(1 - p)$. Obviously, β is much smaller than κ for almost all graphs. In fact one could let $p \to 0$ slowly and SC5 would still be a sure thing. Note that these comments provide the easiest proof that "almost all graphs are hamiltonian," but of course the result is not nearly in the class of Pósa's theorem [Po76], which does it for $p = c(\log n)/n$.

We have seen in Corollary 5.3.1 that the idea of analyzing a theorem probabilistically can also be applied to conjectures. Another good example of this type of result was found by Müller [Mu76], who proved that the famous reconstruction conjecture of Ulam (see Appendix VIII) is almost always true.

Theorem 5.4.1. In Model A with $p = 1/2$, almost all graphs are vertex-reconstructible.

The main part of the proof shows that for any $\varepsilon > 0$ almost all graphs have the property that all subgraphs of order at least $(1 + \varepsilon)n/2$ are mutually nonisomorphic. To do this, nice (small) bounds are given for the expected number of pairs of different subgraphs of order $k > (1 + \varepsilon)n/2$ that are isomorphic. Then this expected value can be shown to have limit zero as $n \to \infty$. But Chinn's theorem (see Appendix VIII) only requires the two-vertex-deleted subgraphs to be mutually nonisomorphic for a graph to be reconstructible.

Lovász [Lo72] used an ingenious application of the principle of inclusion and exclusion to show that a graph of order n is edge-reconstructible if the number q of edges is greater than $\binom{n}{2}/2$. His approach was neatly extended by Müller [Mu77], and the resulting Theorem 5.4.2 demonstrates the strength of the edge-reconstruction conjecture. Before stating the theorem we will introduce some notation and illustrate the way in which the method of inclusion and exclusion will be applied.

Let G_1 and G_2 be graphs of order n with vertex sets V_1 and V_2, respectively. The universal set $\mathscr{U}$ consists of all one-to-one functions from V_1 to V_2. Such a function f is called *monomorphism from G_1 to G_2* if $\{f(u), f(v)\}$ is an edge of G_2 whenever $\{u, v\}$ is an edge of G_1. If H is any graph with vertex set V_1, then we denote by $\langle H, G_2 \rangle$ the number of monomorphisms from H to G_2. Note that if $E(H) = \varnothing$, then $\langle H, G_2 \rangle = n!$.

Now suppose G_1 has q edges denoted by $e_1, e_2, \ldots, e_q$. For each $i = 1$ to q, we define

(5.4.14) $$A_i = \{ f \in \mathscr{U} \mid f(e_i) \in E(G_2) \}.$$

Note that we are sticking to the notation of Appendix VII, and so $S_0 = n!$, the order of the universal set. Next we need an expression for S_r as defined in (7.1) with $r = 1$ to q. Consider any subgraph H of G_1 that has r edges, say

$$E(H) = \{ e_{i_1}, \ldots, e_{i_r} \}.$$

Then it follows from (5.4.14) that

(5.4.15) $$|A_{i_1} \cap \cdots \cap A_{i_r}| = \langle H, G_2 \rangle,$$

since each side of (5.4.15) counts the number of functions in $\mathcal{U}$ that send these r edges to $E(G_2)$. On summing (5.4.15) over all collections of r of the sets A_i, we have

$$(5.4.16) \qquad S_r = \sum \langle H, G_2 \rangle,$$

where the sum on the right is over all subgraphs H of G_1 with exactly r edges.

By definition (7.2), N_0 is the number of functions in $\mathcal{U}$ that send none of the edges of G_1 to the edge set $E(G_2)$. But then these functions must be monomorphisms from G_1 to the complement $\overline{G}_2$, that is,

$$(5.4.17) \qquad N_0 = \langle G_1, \overline{G}_2 \rangle.$$

With this change in notation, the usual form [see (7.7)] of inclusion and exclusion is

$$(5.4.18) \qquad \langle G_1, \overline{G}_2 \rangle = \sum_{i=0}^{q} (-1)^i \sum \langle H, G_2 \rangle,$$

where the inside sum on the right is over all subgraphs H of G_1 that have exactly i edges. Here is a little example to illustrate (5.4.18). Let G_1 and G_2 both be the graph of order 4 with two edges and one isolated vertex. Then $q = 2$, and equation (5.4.18) becomes

$$(5.4.19) \qquad 10 = 4! - 16 + 2.$$

For $k = 0$ to q, N_k is the number of functions in $\mathcal{U}$ that send exactly k edges of G_1 to $E(G_2)$, and the remaining $q - k$ go to $E(\overline{G}_2)$. We define

$$(5.4.20) \qquad \langle G_1, \overline{G}_2 \rangle_k = N_k,$$

and note that when $k = 0$ this notation is consistent with (5.4.17) if the subscript is suppressed.

The general formula (7.8) for inclusion and exclusion takes the form

$$(5.4.21) \qquad \langle G_1, \overline{G}_2 \rangle_k = \sum_{i=0}^{q-k} (-1)^i \binom{i+k}{i} \sum \langle H, G_2 \rangle,$$

where the inside sum on the right is over all subgraphs H of G_1 that have exactly $i + k$ edges. For example, if $k = 1$ and G_1 and G_2 are the same graphs of order 4 used above, this equation becomes

$$12 = 16 - \binom{2}{1} 2.$$

Theorem 5.4.2. Suppose G_1 and G_2 are graphs of order n and size q. Their edge sets are $E(G_1) = \{e_1, \ldots, e_q\}$ and $E(G_2) = \{f_1, \ldots, f_q\}$, and for $i = 1$ to q

$$(5.4.22) \qquad\qquad G_1 - e_i \cong G_2 - f_i.$$

If $q > n \log_2 n$, then G_1 and G_2 are isomorphic.

Proof. First note that condition (5.4.22) in the hypothesis means that G_1 and G_2 have exactly the same subgraphs of size $< q$. We can assume that the vertex sets of G_1 and G_2 are identical so that $\langle H, G_2 \rangle$ counts monomorphisms from H to G_2 whether H is a subgraph of G_1 or of G_2. Then in expression (5.4.16) for S_r, when $r \neq q$, the sum on the right side is the same whether it is taken over subgraphs of size r in G_1 or in G_2.

Now G_2 is substituted for G_1 in equation (5.4.21) for each $k = 0$ to q:

$$(5.4.23) \qquad \langle G_2, \overline{G}_2 \rangle_k = \sum_{i=0}^{q-k} (-1)^i \binom{i+k}{i} \sum \langle H, G_2 \rangle,$$

where the inner sum is over all subgraphs H of G_2 with $i + k$ edges. Next we subtract equation (5.4.21) from (5.4.23).

But the inner sums for corresponding equations are the same whenever the subgraphs H have the same size, and so the cancellation is devastating:

$$(5.4.24)$$
$$\langle G_2, \overline{G}_2 \rangle_k - \langle G_1, \overline{G}_2 \rangle_k = (-1)^{q-k} \binom{q}{k} \langle G_2, G_2 \rangle - (-1)^{q-k} \binom{q}{k} \langle G_1, G_2 \rangle.$$

If we assume that the theorem is false, that is, that G_1 and G_2 are *not* isomorphic, then $\langle G_1, G_2 \rangle = 0$, and these equations are even simpler. On summing the absolute value of both sides of (5.4.24), we have

$$(5.4.25) \quad \sum_{k=0}^{q} |\langle G_2, \overline{G}_2 \rangle_k - \langle G_1, \overline{G}_2 \rangle_k| = \sum_{k=0}^{q} \binom{q}{k} \langle G_2, G_2 \rangle = \langle G_2, G_2 \rangle 2^q.$$

Now we use the fact that $\langle G_2, G_2 \rangle \geq 1$ and apply the triangle inequality to the left side of (5.4.25) to obtain

$$(5.4.26) \qquad\qquad 2^q \leq \sum_{k=0}^{q} \langle G_2, \overline{G}_2 \rangle_k + \sum_{k=0}^{q} \langle G_1, \overline{G}_2 \rangle_k$$
$$\leq n! + n! = n!2.$$

Finally Stirling's formula can be used to show that $q \leq n \log_2 n$, contradicting the hypothesis. Hence $G_1 \cong G_2$. $\qquad \square$

We know from exercise 3.1.2 that in Model A with $pn^2 \to \infty$, almost every graph has at least $(1 - \varepsilon)p\binom{n}{2}$ edges. What should we choose for p so that almost every graph has at least $n \log_2 n$ edges? The answer is

$$(5.4.27) \qquad\qquad p > \frac{2}{(1 - \varepsilon)\log 2} \frac{\log n}{n}.$$

Corollary 5.4.1. In Model A with $p = c(\log n)/n$, if $c > 2/\log 2$, almost every graph is edge-reconstructible.

If G is disconnected, with at least four edges and at least two nontrivial components, then G is edge-reconstructible [BoH77]. From exercise 3.1.3 we know that if $p = c(\log n)/n$ with $c < 1/2$, almost every graph will have isolated edges. We also know that there must be a giant component and therefore almost all of these are edge-reconstructible. It remains to show that Corollary 5.4.1 holds for $1/2 \le c \le 2/\log 2$. In particular, the question arises whether or not the giant component is edge-reconstructible. The corresponding problem in Model B would be to prove that almost all graphs are edge-reconstructible when the number q of edges is in the gap

$$(5.4.28) \qquad\qquad \frac{1}{4}n \log n \le q \le \frac{1}{\log 2}n \log n.$$

We conclude with Stockmeyer's observation that all of these reconstruction results are not significant [St81], because the corresponding results for digraphs are also true and yet the reconstruction conjecture for digraphs is false!

Exercises 5.4

1. Find a simple asymptotic formula for the number $\left(\genfrac{}{}{0pt}{}{\binom{n}{2}}{n-1}\right)$ of labeled graphs of order n and size $q = n - 1$.

2. In Model A with $p = 2c/n$, show that almost all graphs have isolated edges. [Use $E(X)$ and $E(X^2)$ from exercise 3.1.3.]

3. Show for $p = 1/2$ that the expected number of pairs of vertices that do not satisfy Ore's condition SC3 is asymptotic to $n^2/8$.

4. Use the second-moment method to show that in Model A with $p = 1/2$ almost every graph has a pair of vertices that does not satisfy the Ore equation (5.4.7).

5.5. RANDOM FUNCTIONAL DIGRAPHS AND MODEL C

Recall that the sample space Ω of Model C consists of all labeled digraphs of order n in which each vertex has fixed outdegree r. When $r = 1$, these digraphs correspond in an obvious way to functions from a set of n elements into itself. Specifically, if D is a digraph in Ω with $r = 1$ and vertex set $V = \{v_1, \ldots, v_n\}$, then the function $f: V \to V$ that corresponds to D has $f(v_i) = v_j$ whenever v_i is adjacent to v_j. Hence the elements of Ω are often called *functional digraphs*, but note that since digraphs are not permitted to have loops, the functions to which they correspond have no fixed points.

We used a simple proof for Theorem 2.2.1 to show that for $r \geq 2$ almost all digraphs are weakly connected in Model C. The situation is quite different for functional digraphs, as indicated in the following theorem found by Rényi [Re59b] and Katz [Ka55] in their studies on connectivity for functions.

Theorem 5.5.1. In Model C with $r = 1$, the probability that a functional digraph is weakly connected is asymptotic to $[\pi/(2n)]^{1/2} = (1.125\ldots)n^{-1/2}$; hence almost all are disconnected.

Proof. Weakly connected functional digraphs are special orientations of unicyclic graphs and can be formed as follows. Consider a set of $k \geq 2$ rooted, labeled trees with a total of n vertices, orient each edge to form an arc pointing toward the root, and identify the roots with the vertices of an oriented cycle of order k. The formula for the number of sets of k rooted, labeled trees of order n is

$$(5.5.1) \qquad \frac{(n)_k}{(k-1)!} n^{n-k-1},$$

[compare formula (6.3.4) and see exercise 6.3.1], and therefore the number of functional digraphs whose cycle has order k

$$(5.5.2) \qquad (n)_k n^{n-k-1}.$$

On summing the expression (5.5.2) we have the number $C(n)$ of weakly connected functional digraphs:

$$(5.5.3) \qquad C(n) = n^{n-1} \sum_{k=2}^{n} \frac{(n)_k}{n^k}.$$

To estimate the sum in (5.5.3) we use the formulas for $(n)_k/n^k$ in Appendix III. From formula (3.4) we have

$$(5.5.4) \qquad \sum \frac{(n)_k}{n^k} \sim \sum e^{-k^2/2n},$$

where both sums are over all $k \leq n^{2/3}/\omega_n$ and $\omega_n \to \infty$ slowly. But the sum on the right of (5.5.4) can be approximated by an integral if divided by $\sqrt{n}$:

$$(5.5.5) \qquad \sum e^{-k^2/2n} n^{-1/2} \sim \int_0^\infty e^{-y^2/2} \, dy = \left(\frac{\pi}{2}\right)^{1/2}.$$

The exact value of this integral is, of course, well known from the behavior of the normal distribution function (see formulas (1.2) and (1.6) of Chapter VII in [Fe57]). For the upper portion of the sum we use (3.6) of Appendix III to write

$$(5.5.6) \qquad \sum \frac{(n)_k}{n^k} = O(1) \sum e^{-k^2/2n},$$

where the sum is over all k in the interval $n^{2/3}/\omega_n \leq k \leq n$. Then it is easy to see that

$$(5.5.7) \qquad \sum \frac{(n)_k}{n^k} = O(1) n \exp\left(-\frac{n^{1/3}}{2\omega_n^2}\right) = o(1),$$

for the high values of k.

So far, this proves that

$$(5.5.8) \qquad C(n) \sim n^{n-1/2} \left(\frac{\pi}{2}\right)^{1/2}.$$

On dividing by the number $(n-1)^n$ of functional digraphs, the probability of connectivity stated in the theorem is obtained. □

Note that exactly the same asymptotic result (5.5.8) holds if the extra term for $k = 1$, namely, n^{n-1}, is included in the sum on the right side of (5.5.3). Since the number of digraphs in which loops are permitted is n^n, the inclusion or exclusion of loops does not affect the asymptotic proportion of weakly connected digraphs.

Moon [M70] reports other interesting results for random labeled mapping patterns (loops are allowed) such as estimates of the expected number of components, the expected number of vertices in cycles and the expected length of a cycle in a random connected mapping pattern. Meir and Moon [MeM84] have just figured out all of these estimates for the unlabeled case. For example, they found that the probability of weak connectivity for a random unlabeled functional digraph is asymptotic to $(1.128\ldots)n^{-1/2}$.

Now we focus our attention on the main feature of this section, namely, connectivity in Model C with $r \geq 2$. We have already seen in Theorem 2.2.1 that almost all of these digraphs are weakly connected, but Fenner and Frieze found a much stronger result [FeF82]. Recall that we sometimes ignore the orientations of all the arcs of the digraphs in the sample space of Model C and treat them as graphs. First we determine an upper bound for the connectivity of such graphs.

Theorem 5.5.2. In Model C with $r \geq 1$, almost all digraphs have vertices of indegree zero. Hence, when the orientations of their arcs are ignored, almost all have connectivity $\kappa \leq r$.

Proof. Once again to prove the existence of certain types of subgraphs we use the second-moment method. For each digraph D, let $X(D)$ be the number of vertices of indegree zero in D. Then the expectation is

$$E(X) = \binom{n}{1}\left(\frac{\binom{n-2}{r}}{\binom{n-1}{r}}\right)^{n-1} = n\left(\frac{(n-2)_r}{(n-1)_r}\right)^{n-1}$$

$$(5.5.9)$$

$$= n\left(1 - \frac{r}{n-1}\right)^{n-1} \sim \frac{n}{e^r}.$$

Now we write $X = X_1 + \cdots + X_n$, where for $i = 1$ to n, $X_i(D)$ is 1 or 0 according as the vertex v_i has indegree zero or not. Then it can be seen that

$$(5.5.10) \qquad E(X^2) = E(X) + n(n-1)E(X_1 X_2)$$

and

$$E(X_1 X_2) = \left(\frac{\binom{n-2}{r}}{\binom{n-1}{r}}\right)^2\left(\frac{\binom{n-3}{r}}{\binom{n-1}{r}}\right)^{n-2}$$

$$(5.5.11)$$

$$= \left(1 - \frac{r}{n-1}\right)^n\left(1 - \frac{r}{n-2}\right)^{n-2} \sim e^{-2r}.$$

Formula (5.5.9) shows that $E(X) \to \infty$. Then it follows from all three formulas that $E(X)^2 \sim E(X^2)$. Therefore $P(X = 0) \to 0$ (see Proposition 5.5 of Appendix V).   □

Now we need to show that when $r \geq 2$, almost all graphs in Model C have connectivity at least r. We introduce a new random variable X_m that counts the number of sets of m vertices whose deletion results in a disconnected graph. If $E(X_m) \to 0$, then $\kappa \geq m + 1$ for almost all graphs. Therefore we want to show that $E(X_{r-1}) \to 0$. For the expectation we have the upper bound

$$(5.5.12) \qquad E(X_m) \leq \sum \binom{n}{m, k, n - m - k} F(m, k, n),$$

where the sum is over all k such that $k + m \geq r + 1$ and $k \leq \lfloor (n - m)/2 \rfloor$ and

$$(5.5.13) \quad F(m, k, n) = \left(\frac{\binom{m + k - 1}{r}}{\binom{n - 1}{r}} \right)^k \left(\frac{\binom{n - k - 1}{r}}{\binom{n - 1}{r}} \right)^{n - k - m} .$$

The multinomial coefficient in (5.5.12) is the number of ways to select a set R of m vertices to be deleted, a set S of k vertices and another set T of $n - m - k$ vertices. The other factor $F(m, k, n)$ is the sum of the probabilities of all digraphs that have no arc with one vertex in S and the other in T. That is, each of the k vertices in S can only be adjacent to the other $m + k - 1$ vertices in $S \cup R$. And each of the $n - k - m$ vertices in T can only be adjacent to the other $n - k - 1$ vertices in $R \cup T$. The vertices of one digraph might be partitioned in three such sets in several different ways, and so (5.5.12) is not an equality.

With $m \leq r - 1$ and $r + 1 - m \leq k \leq (n - m)/2$, a few simple computations show that

$$(5.5.14) \qquad \binom{n}{m, k, n - m - k} = O(1) \frac{n^n}{k^k (n - m - k)^{n - m - k}}$$

and

$$(5.5.15) \qquad F(m, k, n) = O(1) \left(\frac{m + k}{n} \right)^{kr} \left(\frac{n - k}{n} \right)^{(n - k - m)r} .$$

Now these two formulas are combined, and Stirling's formula is used to show that

$$(5.5.16) \quad \binom{n}{m, k, n - m - k} \cdot$$

$$= O(1) \left(\frac{m + k}{k} \right)^k \left(\frac{n - k}{n - k - m} \right)^{n - k - m}$$

$$\times (m + 1)^{k(r-1)} (n - k)^{(n - k - m)(r - 1)} n^{n(r-1) - mr} .$$

But with $m \leq r - 1$ and k in the required range,

$$(5.5.17) \qquad \left(\frac{m+k}{k}\right)^k \left(\frac{n-k}{n-k-m}\right)^{n-k-m} = O(1).$$

Therefore we have

$$(5.5.18) \qquad E(X_m) = O(1)\sum f(m,k,n)$$

where

$$(5.5.19) \qquad f(m,k,n) = \frac{\left[(m+k)^k(n-k)^{n-k-m}\right]^{r-1}}{n^{n(r-1)-mr}}.$$

It is important to observe (see exercise 5.5.4) that the numerator of $f(m,k,n)$ is a decreasing function of k for $0 \leq k \leq (n-m)/2$. Therefore the first summand in (5.5.18) is the biggest. For the case that concerns us, we have $m = r - 1$ and so $2 \leq k \leq (n - r + 1)/2$. Sheer substitution in (5.5.19) and straightforward simplification then show that

$$(5.5.20) \qquad f(r-1,2,n) = O(1)n^{1-r},$$

and so $f(r-1,2,n) \to 0$ for $r \geq 2$. We also find that

$$(5.5.21) \qquad f(r-1,3,n) = O(1)n^{2(1-r)}.$$

Since $f(r-1,3,n)$ is the biggest of all the remaining terms in the sum and there are less than n such terms, we have

$$(5.5.22) \qquad E(X_{r-1}) = O(1)n^{1-r} + O(1)n^{1+2(1-r)}.$$

Obviously $E(X_{r-1}) \to 0$ as required to prove the following theorem.

Theorem 5.5.3. In Model C with $r \geq 2$, almost all digraphs have the property that at least r vertices must be deleted to produce a digraph that is not weakly connected. Hence, when the orientations of their arcs are ignored, almost all have $\kappa \geq r$.

These two theorems show that in Model C with $r \geq 2$ almost all graphs have $\kappa = r = \delta$, and hence the edge connectivity λ is also equal to r.

Fenner and Frieze discovered a nontrivial characterization of the graphs that arise from the digraphs in the sample space of Model C. They called such graphs *r-orientable*, and found the following relationship between Models B and C.

Theorem 5.5.4. When the number of edges is given by equation (5.1.6) so that in Model B the vertices of degree d are distributed according to Poisson's law, then almost all graphs are d-orientable.

They also created an interesting variation of Model C. In addition to the r vertices adjacent from* each vertex v, also r of the $n - 1$ vertices are selected at random to be adjacent to v. They were able to prove that for $r \geq 2$ almost all of these multidigraphs have the property that r is the minimum number of vertices that must be deleted so that the remaining digraph is not strongly connected. However, it is apparently unknown whether or not for $r = 1$ almost all of these are strongly connected.

Their most exciting recent result for Model C concerns hamiltonicity [FeF83].

Theorem 5.5.5. In Model C with $r \geq 23$, almost all graphs are hamiltonian.

They conjecture that the theorem also holds for r as low as 3. In a related paper [Mc81], McDiarmid found that if r is allowed to grow as a function of n, the required value is about $r = \log n$ to ensure that almost all digraphs in Model C are hamiltonian.

There are only a few other articles concerning Model C. In [ShU82] Shamir and Upfal found a sufficient condition for a 1-factor.

Theorem 5.5.6. In Model C with n even and $r \geq 6$, almost every graph has a 1-factor.

In his study of the expected value of a random assignment problem [Wa79], Walkup used a bipartite version of Model C in which the vertex set V is partitioned in two sets S and T, each of order n. Each vertex of S is adjacent to r vertices of T, and conversely each vertex of T is adjacent to r vertices of S. Therefore there are $\binom{n}{r}^{2r}$ labeled digraphs in the sample space. A *matching* in such a digraph consists of n disjoint arcs, and it is easy to see that the expected number of matchings is

$$(5.5.23) \quad \begin{aligned} E(X) &= n! 2^n \left(\frac{\binom{n-1}{r-1}}{\binom{n}{r}} \right)^n \\ &= n! 2^n \left(\frac{r}{n} \right)^n. \end{aligned}$$

---

*See Appendix VIII, Section on digraphs, for definition of "adjacent from."

If $r = 1$, then $E(X) \to 0$, and so almost no digraph has a matching. But if $r \geq 2$, $E(X) \to \infty$. The second-moment method could be used to show that almost all of these digraphs have a matching. But Walkup [Wa80] found a better way that used a characterization of bipartite graphs with matchings.

Theorem 5.5.7. In the bipartite version of Model C, almost no graph has a matching if $r = 1$, but if $r \geq 2$, almost every graph has a matching.

Exercises 5.5

1. What is the probability that two specified vertices are adjacent in a graph from Model C?

2. Find the expected number of vertices of indegree d in the digraphs of Model C.

3. Derive a simple asymptotic formula for the expectation of the previous exercise [compare formula (5.5.9) where $d = 0$].

4. Show that $y = (m + k)^k (n - k)^{n-k-m}$ is a decreasing function of k for fixed m and n and $0 \leq k \leq (n - m)/2$.

5. What is the expected number of vertices of degree d in a graph from Model C? What is the expected degree of a specified vertex in one of these graphs?

6

RECENT RELATED RESULTS

Nice guys finish last.

Leo Durocher

What other probability models should be studied for graphs? How can we create a graph of specified type at random? How are the vertices of certain graphs distributed with regard to specified properties? These are the questions on which we will focus in this chapter. There are some very interesting recent results that provide answers. From Dixon and Wilf we learn how to generate random graphs. Bender, Canfield and Bollobás have shown how to estimate the number of labeled regular graphs so that these can be used as a sample space. Finally, we sketch Bailey's method for determining the distribution of the vertices in trees by both degree and orbit size simultaneously!

6.1. GENERATION

On the blackboard draw eight vertices $v_1, \ldots, v_8$ evenly spread around a circle of diameter about 2 feet. List the slots available for edges in some convenient order such as $\{v_1, v_2\}, \ldots, \{v_1, v_8\}, \{v_2, v_3\}, \ldots, \{v_7, v_8\}$. For each slot, in order, flip an honest coin.* If the coin shows heads, that slot receives an edge,

---

*The scientific creationist will take 28 coins, flip them all at once and be done with it.

so draw the corresponding chord in the circle. If tails, move on to consider the next slot. On completion of this task, a labeled graph of order 8 will have been generated uniformly at random. That is, each of the 251,548,592 labeled graphs of order 8 has the same chance of being drawn on the board. Still otherwise put, we have used Model A with $p = 1/2$ and $n = 8$. The probability that our random graph is connected is .937..., and it is also extremely likely to be hamiltonian.

For the ultimate in visual display, we find that Read [Re78] has developed a computer program called the Graph Manipulation Package that can be used to show graphs of order ≤ 255 and test them for various properties. It will also produce a random graph of order n and size q. Thus the program works in Model B, where each of these graphs has the same chance of being selected [see (2.1.13)]. The production method has been left for the reader to determine in exercise 6.1.1.

Labeled trees of order n are also easy to generate uniformly at random. The simplest approach relies on Prüfer's elegant proof [Pr18] of Cayley's formula (1.1.7). Prüfer verified this formula by establishing a one-to-one correspondence between labeled trees of order n and all sequences of $n - 2$ positive integers from 1 to n. The number of such sequences is, or course, n^{n-2}. Here's how the correspondence is defined for any tree T of order $n \geq 3$ whose vertices are labeled with the integers from 1 to n:

ALGORITHM 1

STEP 1. Find the vertex v of degree 1 that has the smallest label in T.

STEP 2. Record the label of the vertex adjacent to v, and quit if $n - 2$ labels have been recorded. Otherwise do Step 3.

STEP 3. Delete v from T to form the new tree $T - v$. Return to Step 1 with $T - v$.

Sequences obtained by applying this simple algorithm are often called Prüfer sequences or Prüfer codes. That the algorithm defines a one-to-one correspondence is proved by a routine induction argument on n. Instead of sketching the proof we will provide the inverse algorithm, which finds the unique labeled tree T of order n for a given Prüfer sequence of $n - 2$ elements. Note first that the integers from 1 to n that are missing from the Prüfer sequence must have been precisely the labels on the vertices of degree 1 in T. In general, if a label, say k, occurs exactly m times in the sequence, then the vertex with label k has degree $m + 1$.

ALGORITHM 2

STEP 0. Count the number of elements in the sequence, and add 2 to determine n.

STEP 1. Make a separate list of all the integers from 1 to n that are missing from the Prüfer sequence. Call it the end-vertex list.

STEP 2. Find the smallest label, say k, in the end-vertex list. It must have been the label on the vertex of degree 1 that was deleted to obtain this Prüfer sequence. Suppose m is the first integer in this sequence. Then the vertices with labels k and m must be adjacent in T. Record this edge.

STEP 3. Delete the label k from the end-vertex list. If the Prüfer sequence has only one element m, then there will remain only one element, say j, in the end-vertex list. Then the vertices with labels j and m must also be adjacent. Record that edge and quit. Otherwise delete the first element, m, from the Prüfer sequence. If m is repeated elsewhere in this shortened Prüfer sequence, return to Step 2. Otherwise add m to the end-vertex list before returning to Step 2.

Now it is easy to see how to generate a random labeled tree of order n. Just generate the elements of a Prüfer sequence and apply Algorithm 2.

There is another important consequence of Prüfer's approach. Suppose that, in a given Prüfer sequence, label k occurs $d_k - 1$ times for $k = 1$ to n. For each k the vertex with label k in the associated tree must have degree d_k. Thus $d_1, d_2, \ldots, d_n$ is the degree sequence of a tree of order n, and we can ask for the *total* number of trees in which the vertex with label k has degree d_k for each k. It follows from Prüfer's correspondence that the answer is the number of Prüfer sequences with $d_k - 1$ elements equal to k for $k = 1$ to n. But this number is just the multinomial coefficient

$$(6.1.1) \qquad \binom{n-2}{d_1 - 1, d_2 - 1, \ldots, d_n - 1}.$$

This simple observation has many useful applications (see exercises 6.1.4 and 6.3.1).

The survey of Tinhofer [Ti79] includes methods for generating several other types of labeled graphs at random. We now turn to the more difficult problem of generating unlabeled graphs. As in the labeled case, the generating algorithms are closely related to the enumeration formulas, and so we begin by stating Pólya's formula [H55a] for the number g_n of unlabeled graphs of order

n. We denote partitions of n by vectors $(j) = (j_1, j_2, \ldots, j_n)$, where j_k is the number of parts equal to k, that is,

$$(6.1.2) \qquad \sum_{k=1}^{n} k j_k = n.$$

Each permutation of n objects can be written as a product of disjoint cycles and thus determines a partition of n. The number of different permutations for a given partition (j) is denoted by $h(j)$, and for this we have the well-known formula

$$(6.1.3) \qquad h(j) = \frac{n!}{\prod_{k=1}^{n} k^{j_k} j_k!}.$$

We use (r, s) for the greatest common divisor of r and s. Here is Pólya's formula:

$$(6.1.4) \qquad g_n = \frac{1}{n!} \sum h(j) 2^{g(j)},$$

where the sum is over all partitions of n and

$$(6.1.5) \qquad g(j) = \sum_{k \geq 1} \left\lfloor \frac{k}{2} \right\rfloor j_k + k \binom{j_k}{2} + \sum_{r < s} (r, s) j_r j_s.$$

A detailed proof can be found in Chapter 4 of the book [HP73].

Suppose α is a permutation of the n labels $1, \ldots, n$ and G is a labeled graph of order n. Then αG is the graph whose labels have been permuted by α; that is, for each $i = 1$ to n, the vertex of G with label i has label αi in αG. Let $\Gamma(G)$ be the set of all α such that G and αG are identical labeled graphs. The set $\Gamma(G)$ is the *automorphism group* of G. We sometimes say that G is fixed by α.

Now we can state the algorithm of Dixon and Wilf [DiW83] for generating unlabeled graphs.

ALGORITHM RANDOM GRAPH (UNLABELED)

STEP 1.   Choose a partition $(j) = (j_1, \ldots, j_n)$ of n with probability

$$(6.1.6) \qquad \frac{h(j) 2^{g(j)}}{n! g_n}.$$

STEP 2.   Consider any permutation α with partition (j). Choose uniformly at random a labeled graph G that is fixed by α.

STEP 3.   Ignore the labels in G!

We have not only to justify the steps above but to show how to implement them. First, however, notice that the partition (j) with $j_1 = n$ is to be selected in Step 1 with probability

(6.1.7)
$$\frac{2^{\binom{n}{2}}}{n! g_n}.$$

As mentioned in (1.1.4) of Chapter 1, expression (6.1.7) is asymptotic to 1. Therefore for large n, this partition with n parts equal to 1 is almost certain to be selected in Step 1. Then the permutation of Step 2 is the identity that fixes all labeled graphs. And so Step 2 is finished off by generating a random labeled graph in the manner indicated at the very beginning of this section. So for large n the algorithm almost always produces a random labeled graph whose labels we are to forget. This is exactly what we should expect, since almost all unlabeled graphs have the identity group (see exercise 4.3.1) and therefore each is counted with the same multiplicity, $n!$, among labeled graphs.

The justification of the algorithm rests on a more general method of Dixon and Wilf, which generates a random orbit of a permutation group F with finite object set X. Let the conjugacy classes of F be $F_1, \ldots, F_r$. For any permutation α in F, define the set

(6.1.8)
$$\text{Fix}(\alpha) = \{ x \in X | ax = x \}.$$

The number of orbits of F is denoted by $N(F)$ and can be expressed in terms of $\text{Fix}(\alpha)$ by the following lemma.

Burnside's Lemma*

(6.1.9)
$$N(F) = \frac{1}{|F|} \sum_{\alpha \in F} |\text{Fix}(\alpha)|.$$

The proof is elementary (see Chapter 2 of [HP73]).

---

*It is not clear whose lemma this really is. Group theorists are aware that it was known to many others including Frobenius and even Cauchy. Attributing it to Burnside undoubtedly would have infuriated Frobenius, whose mistrust is shown by the following remarks quoted from the *Old Intelligencer* in a letter to Dedekind dated 7 May 1896: "This is the same Mr. Burnside who, for some years now, has been annoying me by rediscovering every single theorem I have published in group theory. What's more, he doesn't just rediscover them, but rediscovers them without exception in the very same order in which they have been published: first my proof of Sylow's theorem, then the theorem on groups of quadratic order, then the one on groups of order $p^\alpha q$, then the one on groups whose order is a product of 4 or 5 primes, etc., etc. Anyway, it is one of those remarkable and wonderful examples of psychic harmony (*Seelenharmonie*) which can only occur in England or possibly in America."

The next two formulas are also easy to establish. Let $\mathcal{O}$ be any orbit of F. Then

$$(6.1.10) \qquad |F| = |\{(\alpha, x)|x \in \mathcal{O} \cap \text{Fix}(\alpha)\}|.$$

If α and β are in the same conjugacy class of F, then

$$(6.1.11) \qquad |\text{Fix}(\alpha) \cap \mathcal{O}| = |\text{Fix}(\beta) \cap \mathcal{O}|.$$

Finally, we need a system of representatives of the conjugacy classes. For $i = 1$ to r, let α_i be an element of class F_i.

ALGORITHM RANDOM ORBIT

STEP 1.   Choose a conjugacy class F_i of F with probability

$$(6.1.12) \qquad \frac{|F_i||\text{Fix}(\alpha_i)|}{|F|N(F)}$$

STEP 2.   Pick uniformly at random an element x from $\text{Fix}(\alpha_i)$.
STEP 3.   Take the orbit that contains x.

An immediate consequence of Burnside's lemma is that the sum of expression (6.1.12) over all r classes is 1. Formulas (6.1.10) and (6.1.11) can be used to show that each orbit is selected with probability $1/N(F)$. Once class F_i has been chosen in Step 1, the probability of picking an element of a given orbit $\mathcal{O}$ is

$$\frac{|\text{Fix}(\alpha_i) \cap \mathcal{O}|}{|\text{Fix}(\alpha_i)|}.$$

Therefore the overall probability of selecting $\mathcal{O}$ is the sum

$$\sum_{i=1}^{r} \frac{|F_i||\text{Fix}(\alpha_i)|}{|F|N(F)} \frac{|\text{Fix}(\alpha_i) \cap \mathcal{O}|}{|\text{Fix}(\alpha_i)|}$$

$$= \frac{1}{N(F)} \frac{1}{|F|} \sum_{i=1}^{r} \sum_{\alpha \in F_i} |\text{Fix}(\alpha) \cap \mathcal{O}| \qquad \text{by (6.1.11)}$$

$$= \frac{1}{N(F)} \frac{1}{|F|} \sum_{\alpha \in F} |\text{Fix}(\alpha) \cap \mathcal{O}|$$

$$= \frac{1}{N(F)} \qquad \text{by (6.1.10)}$$

It can be seen that Algorithm Random Graph is a special case of Algorithm Random Orbit. The object set X consists of all $2^{\binom{n}{2}}$ labeled graphs of order n. The group F is the representation indicated earlier of the symmetric group S_n, which acts on $G \in X$ by permutation of the labels of G. An unlabeled graph is simply an orbit of F that consists of all the labeled versions of that graph.

To implement Algorithm Random Graph, here are some hints from [DiW83]. For Step 1, list the partitions of n in the order $\lambda_1, \lambda_2, \ldots$, where the number of parts equal to 1 in λ_k is as large as the number in λ_{k+1}. Thus λ_1 has n parts equal to 1 and corresponds to the identity permutation. Let $p_1, p_2, \ldots$ be the probabilities defined by (6.1.6) for these partitions. Pick a random number ξ with $0 \le \xi < 1$. Calculate p_1 and compare it with ξ. If $\xi \le p_1$, select $(j) = \lambda_1$, and go to Step 2. Otherwise calculate p_2, and select $(j) = \lambda_2$ if $\xi \le p_1 + p_2$. Otherwise continue the process as indicated. The expected number of steps to select a partition is seen to be

$$(6.1.13) \qquad \sum_{k=1}^{c(n)} k p_k,$$

where $c(n)$ is the number of partitions (j) of n. Dixon and Wilf estimated this number and found, not surprisingly, that it approaches 1 as $n \to \infty$. Thus for large n we can expect to use the first partition on the list corresponding to the identity permutation. Even for small n they found (6.1.13) to be < 3.

Here is the result of Oberschelp [Ob67] (see also (9.1.24) of [HP73]) used in [DiW83] to estimate (6.1.13). For $k = 0$ to n, define

$$(6.1.14) \qquad g_n^{(k)} = \sum \frac{2^{g(j)} h(j)}{n!},$$

where the sum is over all partitions with $j_1 = n - k$. Note that $g_n^{(1)} = 0$,

$$(6.1.15) \qquad g_n^{(0)} = \frac{2^{\binom{n}{2}}}{n!}$$

and

$$(6.1.16) \qquad g_n = \sum_{k=0}^{n} g_n^{(k)}.$$

Oberschelp proved that for $r = 0$ to n,

$$(6.1.17) \qquad \sum_{k=r}^{n} g_n^{(k)} = g_n^{(0)} O(n^r 2^{-rn/2}).$$

From this formula with $r = 2$, it follows that (6.1.13) is bounded by

(6.1.18)
$$\frac{2^{\binom{n}{2}}}{n!g_n} + c(n)O(n^2 2^{-n}).$$

Since the number of partitions $c(n)$ behaves like $ae^{b\sqrt{n}}/n$, for constants a and b, expression (6.1.18) approaches 1 as $n \to \infty$.

As for Step 2, suppose (j) is the partition selected by Step 1 and α is its corresponding permutation of the labels $1, \ldots, n$. In deriving formula (6.1.4) from Burnside's lemma, one observers that α induces a permutation α^* of the 2-subsets of $\{1, \ldots, n\}$:

(6.1.19)
$$\alpha^*\{i, j\} = \{\alpha i, \alpha j\}.$$

As a permutation of the $\binom{n}{2}$ 2-subsets, α^* consists of disjoint cycles of 2-subsets. Given one of these cycles, each 2-subset of it must be included in a graph G fixed by α, or none can be included. The total number of these cycles is $g(j)$ given in formula (6.1.5). Therefore the number of graphs fixed by α is $2^{g(j)}$. To handle Step 2, we run through a list of these $g(j)$ cycles, and for each one we select all of its edges (or none of them) as edges of G with probability $1/2$. Then each of the $2^{g(j)}$ graphs fixed by α has the same chance of being chosen. Certainly this approach of Dixon and Wilf can be applied to many other types of structures.

The generation of unlabeled trees by Wilf [Wi81] takes its cue from an idea of Cayley that enumerates trees by number of centroidal vertices [C81]. First, of course, Cayley counted *rooted* trees in which one vertex is distinguished from the others. Similarly, to generate unlabeled trees one needs to be able to generate the rooted ones. And, as usual, generation is closely linked with enumeration.

Here is the recurrence relation for the number r_n of unlabeled rooted trees of order n. Let $r_1 = 1$, and for $n > 1$,

(6.1.20)
$$r_n = \frac{1}{n-1} \sum_{k=1}^{n-1} r_{n-k} \left(\sum_{d|k} dr_d \right).$$

Cayley used this relation to calculate r_n for $n \leq 13$. Its justification rests on Theorem 6.3.1 (see exercise 6.3.2).

The following algorithm for generating a random rooted tree of order n was derived by Nijenhuis and Wilf [NiW78]. It is recursive, and so we assume that we can generate rooted trees of order $k < n$. It also requires the computation [from (6.1.20)] of r_k for $k = 1$ to n.

ALGORITHM RANDOM ROOTED TREE

STEP 1.   Select a pair of integers j, d, where $1 \leq jd \leq n - 1$ with probability

(6.1.21)
$$\frac{r_{n-jd}\, dr_d}{(n-1)r_n}.$$

STEP 2.   Pick a rooted tree T' of order $n - jd$ with probability $1/r_{n-jd}$. Pick another one T'' of order d with probability $1/r_d$.

STEP 3.   Consider j copies of T'', and join each of the j roots with an edge to the root v of T'. Take the resulting tree T of order n with root v.

Now we check that each rooted tree T has the same chance of being constructed, namely $1/r_n$. From the algorithm it follows that this probability is

(6.1.22)
$$\sum \frac{r_{n-jd}\, dr_d}{(n-r)r_n} \frac{1}{r_{n-jd}} \frac{1}{r_d}$$

summed over all j, d that produce the given tree T. Suppose T is rooted at v. Denote by $T - v$ the subgraph obtained by removing from T the vertex v and all its incident edges. Then $T - v$ forms a set of trees with each one rooted at the vertex that was adjacent to v. Suppose further that the nonisomorphic rooted trees of $T - v$ are $H_1, \ldots, H_m$. For $i = 1$ to m, let d_i be the order of H_i and μ_i the number of copies of it in $T - v$. Then (6.1.22) is seen to be

(6.1.23)
$$\sum_{i=1}^{m} \sum_{j=1}^{\mu_i} \frac{d_i}{(n-1)r_n},$$

which in turn simplifies to $1/r_n$.

Given a tree T with vertex v, the *weight* of a vertex v is the maximum order of the components of $T - v$. The *centroid* of T consists of the vertices of minimum weight. The concept was introduced by Jordan (see p. 48 of [BiLW76]), and the following elementary observations were used by Cayley.

(i)   The centroid has just one vertex or two adjacent vertices.

(ii)   In a tree of order n with just one centroidal vertex v, the weight of v is $\leq (n-1)/2$. These are all formed by joining with an edge the roots of trees of order $\leq (n-1)/2$ to a new root.

(iii)   A tree with two centroidal vertices has even order n, and they are formed by joining with an edge the roots of two rooted trees of order $n/2$.

Cayley used these properties together with the numbers of rooted trees to calculate the numbers $t_n^{(1)}$ and $t_n^{(2)}$ of trees of order n with one and two centroidal vertices, respectively. Then the number t_n of trees of order n is just $t_n^{(1)} + t_n^{(2)}$. From property (iii) it follows immediately that trees of order n with a double centroid are formed by selecting two rooted trees of order $n/2$ with repetition. Thus for n odd, $t_n^{(2)} = 0$, but for n even,

$$(6.1.24) \qquad t_n^{(2)} = \binom{r_{n/2} + 1}{2}.$$

For single centroids Cayley used generating functions, observing that $t_n^{(1)}$ is the coefficient of x^{n-1} in

$$(6.1.25) \qquad \prod_{k=1}^{(n-1)/2} (1 - x^k)^{-r_k}.$$

This conclusion is a direct consequence of property (ii).

To generate a random tree we first formulate separate algorithms for generating trees with single and double centroids.

ALGORITHM DOUBLE CENTROID

STEP 1. With probability $(1 + r_{n/2})^{-1}$, pick a rooted tree of order $n/2$ at random. Make two photocopies, and join the roots in the two copies by an edge to form a tree of order n with a symmetry line. The two roots constitute the double centroid.

STEP 2. We are here with probability $r_{n/2}/(1 + r_{n/2})$. Pick a rooted tree T' of order $n/2$, and choose another T'' with replacement. These trees are identical with probability $1/r_{n/2}$. Join the roots of T' and T'' to form a tree with a double centroid.

Property (ii) implies that to generate a tree with a single centroid, we need only create a set of rooted trees each of order $\leq (n - 1)/2$ but with a total of $n - 1$ vertices. Let f_i be the number of these sets of rooted trees of total order i. The corresponding generating function is

$$(6.1.26) \qquad f(x) = \sum_{i=0}^{\infty} f_i x^i,$$

where $f_0 = 1$. Thus $f(x)$ is just another name for (6.1.25):

$$(6.1.27) \qquad f(x) = \prod_{k=1}^{(n-1)/2} (1 - x^k)^{-r_k}.$$

A recurrence relation for the coefficients is easy to derive (see exercise 6.1.6):

$$(6.1.28) \qquad mf_m = \sum_{j=1} \sum_{d=1} f_{m-jd} \, dr_d.$$

where $1 \le jd \le m$ and $d \le (n-1)/2$. Note that the first two steps of the algorithm are recursive.

ALGORITHM SINGLE CENTROID

STEP 1.   Set $m = n - 1$, and select a pair of integers j, d, where $1 \le jd \le m$ and $d \le (n-1)/2$ with probability

$$(6.1.29) \qquad \frac{f_{n-jd} \, dr_d}{mf_m}.$$

STEP 2.   Generate a set of rooted trees each of order $\le (n-1)/2$ but with a total of $m - jd$ vertices. Generate also a rooted tree of order d, and make j copies.

STEP 3.   Join by edges the roots of each of these j copies and the roots of the other set to a new vertex v. This new tree has order n and single centroid v.

The algorithm is justified in much the same way as Algorithm Random Tree. And the two centroid algorithms are easily coordinated to produce unrooted trees. If n is odd, just use the single-centroid algorithm. If n is even, use the single-centroid algorithm with probability $t_n^{(1)}/t_n$ and the double with probability $t_n^{(2)}/t_n$.

Exercises 6.1

1.   Assume that your computer generates random numbers between 0 and 1. Sketch an efficient procedure for producing uniformly at random a k-subset of a set of n elements.

2.   Draw the labeled tree whose Prüfer sequence is $4, 8, 2, 2, 3, 2, 4$.

3.   What is the order of magnitude of the number of steps required to find the labeled tree of order n for a given Prüfer sequence?

4.   How many labeled trees of order n have vertices only of degree 1 or 3? Sketch a procedure for generating such a tree at random.

5.   Approximately how many elementary operations are required to compute the number of rooted trees from (6.1.20)? How many for the rooted tree algorithm?

6.   Carry out the steps required to derive the recurrence relation (6.1.28) for f_m.

6.2.   RANDOM REGULAR GRAPHS

In a *regular* graph each vertex has the same degree. When that degree is r, the graph is called *r-regular*. When the degree is 3, the graph is *cubic*. Here's how Redfield would have generated a random cubic.

Take $2n$ triangles on $6n$ labeled vertices. Drop $3n$ mutually disjoint edges at random on the triangles with vertex falling on vertex in a one-to-one correspondence. Now shrink each triangle to a single vertex, leaving only the fallen edges visible. The result may be a labeled cubic graph of order $2n$, but loops or multiple edges could creep in. To generate a cubic graph, just keep bombing the triangles until a direct hit occurs.

Redfield [R27] made a rather thorough investigation of the enumeration of structures obtained by stacking up several copies of graphs of the same order, and he found many applications. But his extraordinary paper went unnoticed for thirty years. In the meantime, R. C. Read independently rediscovered Redfield's main theorem, called the Superposition Theorem in [Re59], and many new applications. In particular, Read found a clever variation of the Superposition Theorem that led to formulas for the number of regular graphs of given order, and loops and/or multiple edges could be included or excluded as desired. Furthermore, the numbers are all neatly expressed in terms of the cycle indexes of wreath products of the symmetric and alternating groups. But computation from such formulas is extremely awkward, and Read treated only the cubic case. Nevertheless he found the asymptotic behavior of the numbers $G_{2n}^{(3)}$ of labeled cubics of order $2n$:

(6.2.1)
$$G_{2n}^{(3)} \sim \frac{(6n)!e^{-2}}{(3!)^{2n}2^{3n}(3n)!}.$$

Read later derived a recurrence relation [Re70] for counting labeled cubics. Its verification was nicely simplified by Wormald [Wo79b], using the idea of enumerating graphs obtained from cubics by deleting a vertex. Read and Wormald combined forces in [ReW80], where they were able to extend their reduction methods to cover labeled 4-regular graphs. Their solution is efficient and was used to calculate the number of 4-regular graphs of order ≤ 13. The corresponding unlabeled problem is quite a bit more complicated. There is a brief outline of Robinson's solution in [Ro77b].

As for asymptotics, Wormald used in his thesis [Wo78c] a result of A. Békéssy, P. Békéssy and Komlós [BeBK72] to generalize Read's formula (6.2.1). Here is his result.

Theorem 6.2.1. Let $G_n^{(r)}$ be the number of r-regular labeled graphs of order n and size $q = nr/2$. Then for fixed $r \geq 1$,

$$(6.2.2) \qquad G_n^{(r)} \sim \frac{(rn)!e^{(1-r^2)/4}}{(r!)^n 2^q q!}.$$

It is assumed that in (6.2.2) nr is even as $n \to \infty$.

At about the same time, Bender and Canfield [BeC78] independently developed a different method for estimating the number of labeled graphs with a given degree sequence. Theorem 6.2.1 is a special case of their general result. Next we shall sketch an entirely different proof due to Bollobás (see [Bo80] or [Bo81a]). It is preferred here because of its probabilistic nature and Redfieldian flavor and most of all for its simplicity and applicability.

Proof of Bollobás. When $r = 1$, there is only one unlabeled graph whose vertices have degree 1, namely, the graph consisting of $q = n/2$ isolated edges. Its automorphism group has order $2^q q!$, and so the number of ways in which it can be labeled according to (1.1.1) is

$$(6.2.3) \qquad \frac{n!}{2^q q!}.$$

Therefore (6.2.2) is an equality for $r = 1$.

For $r \geq 2$, we begin by defining a configuration. Let

$$(6.2.4) \qquad V = \bigcup_{i=1}^{n} V_i$$

be a disjoint union of rn labeled vertices with $|V_i| = r$ for each $i = 1$ to n. A *configuration*, denoted by F, is just a 1-regular graph with vertex set V. Let $\Phi = \Phi(n, r)$ be the set of configurations. Since a configuration consists of $q = rn/2$ isolated edges, we see, just as in the case above for $r = 1$, that the number of configurations is

$$(6.2.5) \qquad |\Phi(n, r)| = \frac{(rn)!}{2^q q!}.$$

If we shrink each set V_i of a configuration F to a single vertex v_i, loops or multiple edges can be formed. Therefore we define a special subclass of *good* configurations for which this does not happen. Let $\Omega = \Omega(n, r)$ be the set of configurations F in Φ that satisfy the following two properties:

(i)   No edge of F has its vertices in the same set V_i.
(ii) No pair of edges of F have their vertices in the union of two of the sets V_i.

These good configurations correspond to r-regular graphs of order n. This correspondence is many-to-one, as the next equation shows. Here the set of labeled r-regular graphs of order n is denoted by $\mathscr{G}_n^{(r)}$:

$$(6.2.6) \qquad |\Omega(n, r)| = (r!)^n |\mathscr{G}_n^{(r)}|.$$

The proof can be finished by finding the proportion $|\Omega|/|\Phi|$ of good configurations. Therefore we regard the set Φ of configurations as a sample space in which each configuration F has the same probability:

$$(6.2.7) \qquad P(F) = \frac{2^q q!}{(rn)!}.$$

For each $l \geq 1$, Bollobás defined an *l-cycle* of a configuration F to be a set of l edges in F, say $e_1, \ldots, e_l$, such that for some set of l distinct sets $V_{i_1}, \ldots, V_{i_l}$ the edge e_k has one vertex in V_{i_k} and the other in $V_{i_{k+1}}$, with $V_{i_{l+1}} = V_{i_1}$. Hence a 1-cycle is an edge with both vertices in the same set V_i, a violation of property (i). And if F has a 2-cycle, property (ii) is not satisfied. For $l \geq 1$, define the random variables $Y_l(F)$ to be the number of l-cycles of F. Then, the good configurations F are precisely those for which $Y_1(F) = Y_2(F) = 0$.

A straightforward computation (see exercise 6.2.2) shows that

$$(6.2.8) \qquad E(Y_l) \sim \frac{(r-1)^l}{2l} = \lambda_l$$

for each $l \geq 1$. (See also [Wo8ld].)

Now we let

$$(6.2.9) \qquad X = Y_1 + \cdots + Y_k$$

and apply the Bonferonni inequalities, in particular, equation (5.17) of Appendix V. We need S_j, which is given by (5.13), and so we let $E(s_1, \ldots, s_k)$ be the expected number of j-sets of s_1 1-cycles, $\ldots$, s_k k-cycles, where $j = s_1$

$+ \cdots + s_k$. Then we have

$$(6.2.10) \qquad S_j = \sum_1 E(s_1, \ldots, s_k),$$

where the sum $\sum_1$ is over all k-sequences of nonnegative integers and $j = s_1 + \cdots + s_k$. Now, for example, the principle of inclusion and exclusion takes the form

$$(6.2.11) \qquad P(X = 0) = \sum_{j \geq 0} (-1)^j \sum_1 E(s_1, \ldots, s_k).$$

If we can show that

$$(6.2.12) \qquad E(s_1, \ldots, s_k) \sim \prod_{l=1}^{k} \frac{(\lambda_l)^{s_l}}{s_l!},$$

then it follows from the Bonferonni inequalities that

$$(6.2.13) \qquad P(X = 0) \rightarrow \prod_{l=1}^{k} e^{-\lambda_l},$$

as $n \rightarrow \infty$, and we say that the random variables $Y_1, \ldots, Y_k$ are distributed in the limit as k independent Poisson random variables.

The verification of (6.2.12) begins with a formula for $E(s_1, \ldots, s_k)$. Let t be the total number of edges in the $j = s_1 + \cdots + s_k$ cycles; hence

$$(6.2.14) \qquad t = \sum_{l=1}^{k} l s_l.$$

Suppose $m \leq t$ is the number of sets V_i chosen for the vertices of the j cycles and $f(m)$ denotes the number of choices for their t edges. Then the formula is

$$(6.2.15) \qquad E(s_1, \ldots, s_k) = P(F) \frac{(2(q-t))!}{2^{q-t}(q-t)!} \sum_{m=1}^{t} \binom{n}{m} f(m),$$

where $P(F)$ is the probability of a configuration F given by (6.2.7) and the next factor is just the number of ways of dropping the remaining $q - t$ edges on the other $2(q - t)$ vertices.

A bit of work shows that

$$(6.2.16) \qquad \binom{n}{t} f(t) = \frac{(n)_t [r(r-1)]^t}{\displaystyle\prod_{l=1}^{k} (2l)^{s_l} s_l!}.$$

From this it follows that the top term on the right side of (6.2.15), denoted by E_1, is

(6.2.17)
$$E_1 = \frac{(q)_t (2r)'(n)_t}{(2q)_{2t}} \prod_{l=1}^{k} \frac{(\lambda_l)^{s_l}}{s_l!}.$$

Now it is easy to show that E_1 is asymptotic to the right side of (6.2.12).

Denote by E_2 the contribution of the right side of (6.2.15) for all $m \le t - 1$. Note that these are all the terms for which there is at least one set V_i containing vertices of different cycles in the set of j cycles. For any $m \le t$ we have

(6.2.18)
$$f(m) \le (rm)_{2t} \le (rm)^{2t}.$$

Using this bound in (6.2.15), we find

(6.2.19)
$$E_2 \le \frac{(q)_t 2^t}{(2g)_{2t}} \sum_{m=1}^{t-1} (n)_m (rm)^{2t}.$$

Then it easily follows that

(6.2.20)
$$E_2 = O\left(\frac{1}{n}\right) = o(1),$$

and so (6.2.13) is established. To prove the theorem we only need (6.2.13) for $k = 2$:

(6.2.21)
$$P(Y_1 + Y_2 = 0) \to e^{-\lambda_1 - \lambda_2}.$$

But from (6.2.8),

(6.2.22)
$$\lambda_1 + \lambda_2 = \frac{r^2 - 1}{4},$$

and (6.2.2) follows from these last two equations and (6.2.5) and (6.2.6).   □

Now that there is a good estimate for the number of regular graphs of large order, it is practical to use $\mathcal{G}_n^{(r)}$ as a sample space with the equiprobable model. So, for example, the probability that a regular graph is connected is just the proportion of those of order n that are connected. Bollobás [Bo81a] used the configuration model described in his proof above to obtain the following corollary.

Corollary 6.2.1. For fixed $r \ge 3$, almost all r-regular graphs are r-connected, that is, r vertices must be removed to disconnect the graph.

This corollary was also found independently by Wormald [Wo81c]. It shows that almost all r-regular graphs have $\kappa = \lambda = r$. From the corollary of Tutte's theorem in Appendix VIII, we can conclude that almost all r-regular graphs have a 1-factor. No doubt this statement could be considerably strengthened.

There are a few further results in [Bo81a] on random regular graphs and a detailed study of the diameter by Bollobás and de la Vega [BoV82]. Wormald has also made some contributions to the problem of generating random regular graphs [Wo84].

Do almost all regular graphs have the identity automorphism group? The answer has just been found to be yes. Let $g_n^{(r)}$ be the number of unlabeled r-regular graphs of order n.

Theorem 6.2.2. Almost all r-regular graphs have the identity automorphism group, that is,

$$(6.2.23) \qquad\qquad g_n^{(r)} \sim \frac{G_n^{(r)}}{n!}.$$

Details are in an article of McKay and Wormald [McW-U] or the paper of Bollobás [Bo82c], where the configuration approach is nicely exploited once more.

It remains to settle the most intriguing question about random regular graphs. Are almost all r-regular graphs hamiltonian? Bollobás [Bo83] used his configuration model to find that the answer is yes provided r is about 10 million. The best result so far has been established by Fenner and Frieze [FeF-U] and is stated in the next theorem.

Theorem 6.2.3. There is a number r_0, $3 \le r_0 \le 796$, such that for fixed $r \ge r_0$ almost every r-regular graph is hamiltonian.

The prevailing gossip favors the opinion that the upper bound on r_0 can be reduced to 3. And there is some nontrivial evidence. For example, Robinson and Wormald [RoW84] have found that almost all bipartite cubics are hamiltonian. They also showed that 98.4% of all large cubics are hamiltonian!

Exercises 6.2

1. In the process described at the beginning of this section, where $3n$ edges are dropped at random on the labeled vertices of $2n$ triangles, what is the asymptotic probability of a direct hit? That is, a genuine cubic graph (no loops, no multiple edges, no special features at all!) is obtained.

2. Find the exact formula for the expected number of l-cycles, and verify (6.2.8).

3. Verify (6.2.16), and show that E_1 in (6.2.17) is asymptotic to the right side of (6.2.12).

4. Prove that (6.2.19) implies (6.2.20).

5. If you've never proved a Bonferonni inequality such as (5.17), isn't it about time? Go for it.

6.3. RANDOM TREES

Pick any vertex of any labeled tree of order n. What are the changes that it has degree 1? The answer for large n approaches $1/e$; same answer as the *problème des rencontres*! It's easy to establish this early result on random trees of Rényi [Re59a] by a little counting argument. The sample space Ω consists of all the vertices of the labeled trees of order n and therefore by Cayley's theorem (1.1.7) has order n^{n-1}. We use the equiprobable model, where each vertex v in Ω has the same probability $1/n^{n-1}$. The random variable X is defined by $X(v) = \deg v$ for all vertices v in Ω. Then the probability that a vertex has degree 1 is

$$(6.3.1) \qquad P(X = 1) = \frac{\binom{n}{1, 1, n-2}(n-1)^{n-3}}{n^{n-1}},$$

where the multinomial coefficient is the number of ways to select a label for the vertex of degree 1 and another for the adjacent vertex, and the other factor of the numerator is the number of labeled trees of order $n - 1$. It follows easily that if $n \to \infty$, then

$$(6.3.2) \qquad P(X = 1) \to \frac{1}{e}.$$

But it is not quite so obvious that for any fixed $k \geq 1$

$$(6.3.3) \qquad P(X = k) \to \frac{1}{(k-1)!e}.$$

This last result (see exercise 6.3.1) can be derived from the following formula of Clarke [Cl58] for the number of vertices in Ω of degree k:

$$(6.3.4) \qquad n\binom{n-1}{k}k(n-1)^{n-k-2}.$$

For more details about this problem and many other important results on random labeled trees, the reader may consult Chapter 7 of Moon's excellent monograph [M67]. We shall now turn to the same problem for unlabeled trees. For this we require several definitions and a summary of background material.

The generating functions for rooted trees and trees are denoted by $T(x)$ and $t(x)$, respectively. That is,

$$(6.3.5) \qquad\qquad T(x) = \sum_{n=1}^{\infty} r_n x^n,$$

where r_n is the number of rooted trees of order n, and

$$(6.3.6) \qquad\qquad t(x) = \sum_{n=1}^{\infty} t_n x^n,$$

where t_n is the number of trees of order n (unlabeled, of course!). Sometimes we say that "$T(x)$ counts rooted trees" or "$t(x)$ counts trees."

Cayley [C57] expressed $T(x)$ as an infinite product from which the recurrence relation (6.1.20) follows.

Theorem 6.3.1. The generating function $T(x)$ for rooted trees is given by

$$(6.3.7) \qquad\qquad T(x) = x \prod_{k=1}^{\infty} (1 - x^k)^{-r_k}.$$

Proof. If T is a tree rooted at vertex v, a *branch of T at v* is a component of $T - v$. The branches of T can be regarded as trees rooted at the vertex adjacent to v. A typical term in the product of (6.3.7) is $(1 - x^k)^{-1}$. This term can be interpreted as the generating function for one particular branch of order k *with repetition*. The product of $(1 - x^k)^{-1}$ over all rooted trees of order $k \geq 1$ counts all possible combinations of branches, and multiplication by x introduces the contribution of the root itself. This establishes (6.3.7). $\qquad\square$

We have already indicated in Section 6.1 how Cayley used the centroid to count trees. A much better method was introduced by Otter [O48], who expressed $t(x)$ in terms of $T(x)$.

For any tree T, $n^*(T)$ is the number of ways T can be rooted at a vertex, that is, the number of orbits of vertices determined by the automorphism group of T. Similarly, $q^*(T)$ is the number of edge-rooted versions of T. If T has an edge whose vertices can be interchanged by an automorphism, the edge is called a *symmetry edge* and $s(T)$ denotes the number of these. Hence $s(T) = 0$ or 1.

Otter's Lemma. For any tree T,

(6.3.8) $$1 = n^*(T) - q^*(T) + s(T).$$

Proof. To verify (6.3.8), consider three cases. First suppose T has a single central vertex w. Then $s(T) = 0$. Now match each vertex $v \neq w$ with the first edge of the path from v to w. This correspondence also matches vertex-rooted trees with edge-rooted trees except for the central vertex w. Hence $n^*(T)$ exceeds $q^*(T)$ by 1.

If T has two central vertices, treat the cases $s(T) = 0$ and $s(T) = 1$ separately and use the same correspondence indicated above.   □

The lemma is easily changed to a statement about generating functions for trees. Just multiply (6.3.8) by x^n and sum the equation over all trees T of order $n \geq 1$. The sum of the left side is $t(x)$, the generating function for trees.

On the right side, $n^*(T)x^n$ sums to $T(x)$, the series for rooted trees. Trees rooted at an edge are formed by joining with an edge the roots of two vertex-rooted trees. The square $T(x)^2$ counts *ordered* pairs of rooted trees, and $T(x^2)$ counts pairs of duplicates. Therefore $[T(x)^2 + T(x^2)]/2$ counts unordered pairs, and $T(x^2)$ handles those with a symmetry edge. These observations prove Otter's formula for trees.

Theorem 6.3.2. The generating function $t(x)$ for trees is given by

(6.3.9) $$t(x) = T(x) - \tfrac{1}{2}\left[T(x)^2 + T(x^2)\right] + T(x^2).$$

The great advantage of this approach of Otter will be seen when we study other properties of trees. Relations are easy to derive, and proofs are clean and swift.

The derivation of Otter's asymptotic results is quite a bit more complicated. Explanations are available in several sources such as [O48] or Chapter 9 of [HP73], but the most detailed account is the article of Harary, Robinson and Schwenk [HRS75], where the whole process is broken down into 20 convenient steps that can be applied to other types of trees. Several of these steps show that the series $T(x)$ has positive radius of convergence $\rho = .3383219\ldots$ and in fact $T(\rho) = 1$. We use the symbol $\sim$ with power series as well as sequences. For example, $T(x) \sim a_n$ means that a_n and the nth coefficient of $T(x)$ are asymptotically equivalent. We only need the following statement of Otter's asymptotic theorems.

Theorem 6.3.3. The asymptotic behavior of the numbers of rooted and unrooted trees is given by

$$(6.3.10) \qquad T(x) \sim \frac{.4399237\ldots}{n^{3/2}\rho^n}$$

and

$$(6.3.11) \qquad t(x) \sim \frac{.5349485\ldots}{n^{5/2}\rho^n}.$$

To determine the probability of a vertex of degree 1 in an unlabeled tree, Robinson and Schwenk [RoS75] defined the generating function $D(x)$, which has as the coefficient of x^n for $n \geq 1$ the number of vertices of degree 1 (excluding the root) in all trees of order $n + 1$ that are rooted at a vertex of degree 1. Thus $D(x)$ begins

$$D(x) = x + x^2 + 3x^3 + \cdots.$$

They found that $D(x)$ satisfies the relation

$$(6.3.12) \qquad D(x) = x + T(x) \sum_{i=1}^{\infty} D(x^i).$$

To see this, suppose T is a tree of order $n + 1 \geq 3$ rooted at a vertex v of degree 1 that is adjacent to vertex u. Suppose one of the branches of $T - v$ at v appears with multiplicity k. The vertices of degree 1 in these k branches are enumerated by

$$T(x) \sum_{i=1}^{k} D(x^i).$$

Let $d(x)$ count vertices of degree 1 in unrooted trees. This generating function is easily expressed in terms of $T(x)$ and $D(x)$ when one recalls the manner in which Otter derived (6.3.9). The result is

$$(6.3.13) \quad d(x) = \{ D(x) - x + xT(x) \} - T(x)D(x) + D(x^2).$$

The reasoning is as follows. First, $\{ D(x) - x + xT(x) \}$ counts vertices of degree 1 in trees rooted at a vertex, with $xT(x)$ counting the root whenever it has degree 1 and $D(x) - x$ counting all the rest. Next $T(x)D(x) + D(x^2)$ counts vertices of degree 1 in trees rooted at an edge, while $2D(x^2)$ counts them in trees with a symmetry edge.

Now an important substitution is made. Replace the first occurrence of $D(x)$ on the right side of (6.3.13) by the right side of (6.3.12). We then have

$$(6.3.14) \qquad d(x) = T(x)\left\{x + \sum_{i=2}^{\infty} D(x^i)\right\} + D(x^2).$$

Equations (6.3.12) and (6.3.13) imply that both $D(x)$ and $d(x)$ have the same radius of convergence as $T(x)$, namely, ρ. Therefore the function

$$(6.3.15) \qquad \left\{x + \sum_{i=2}^{\infty} D(x^i)\right\}$$

can be shown to be analytic at $x = \rho$, and so

$$(6.3.16) \qquad d(x) \sim T(x)\left\{\rho + \sum_{i=2}^{\infty} D(\rho^i)\right\}.$$

Robinson and Schwenk calculated the first 36 coefficients of $D(x)$ and estimated the expression (6.3.15) at $x = \rho$. Then, by combining (6.3.10), (6.3.11) and (6.3.16), they obtained the proportion of vertices of degree 1.

Theorem 6.3.4. The asymptotic probability of a vertex of degree 1 in trees of order n is

$$(6.3.17) \qquad \frac{r_n}{nt_n}\left\{\rho + \sum_{i=2}^{\infty} D(\rho^i)\right\} \sim .438156\ldots .$$

So about 44% of the vertices of a large tree can be expected to have degree 1. Of course, Robinson and Schwenk treated the cases for degree $r \geq 1$, and their results are in the first column of Table 6.3.1.

TABLE 6.3.1. Percentages of Vertices by Degree and by Orbit Size in a Large Unlabeled Random Tree

	Degree	Orbit Size
1	43.8156	69.9539
2	29.3998	17.3626
3	15.9114	6.9888
4	6.8592	3.5219
5	2.6027	1.2204
6	0.9259	0.6187
7	0.3198	0.1935
Total	99.8344	99.8598

In [HP79] we sought the probability that a vertex of a tree is fixed by its automorphism group. Let $F(x)$ count fixed vertices in rooted trees (including the root, which is always fixed by the group of a rooted tree), and let $f(x)$ count fixed vertices in trees. For rooted trees we found

$$(6.3.18) \qquad F(x) = T(x) + T(x)F(x) - T(x)F(x^2)$$

from which the coefficients of $F(x)$ can be calculated. To verify this relation, first note that $T(x)$ counts the roots, which are always fixed. The next term $T(x)F(x)$ counts the fixed vertices in each branch of a rooted tree. But it includes too much, counting as fixed the vertices of a branch even when that branch occurs more than once. Subtracting $T(x)F(x^2)$ neatly compensates for this overdose.

In the unrooted case we obtained

$$(6.3.19) \qquad f(x) = F(x) - T(x)F(x) - F(x^2).$$

This formula is also quickly verified using the Otter approach. Just observe that fixed vertices in trees rooted at an edge are counted by $T(x)F(x) + F(x^2)$, but there are no fixed vertices in a tree with a symmetry edge.

As before, we substitute the right side of (6.3.18) for the first occurrence of $F(x)$ on the right of (6.3.19), and we have

$$(6.3.20) \qquad f(x) = T(x)\big[1 - F(x^2)\big] - F(x^2).$$

Since $F(x^2)$ is analytic at $x = \rho$,

$$(6.3.21) \qquad f(x) \sim T(x)\big[1 - F(\rho^2)\big],$$

and the theorem follows.

Theorem 6.3.5. The asymptotic probability of a fixed vertex in trees of order n is

$$(6.3.22) \qquad \frac{r_n}{nt_n}\big\{1 - F(\rho^2)\big\} \sim .6995\ldots.$$

Thus about 70% of the vertices of a large tree can be expected to be fixed by the automorphism group. These vertices form a fixed subtree that spreads out from the center. The percentages for vertices in higher orbit sizes are given in the second column of Table 6.3.1.

Bailey [Ba82] found the ultimate refinement of Theorems 6.3.4 and 6.3.5 by determining the proportions of vertices of degree r and orbit size s. To state his results, we use S_k to denote the symmetric group on k objects and $Z(S_k) = Z(S_k; s_1, s_2, \ldots)$ for its cycle index (see Chapter 2 of [HP73]). For any generating function $g(x)$ we denote by

$$Z(S_k)[s_i \rightarrow g(x^i)]$$

the result of substituting $g(x^i)$ for each variable s_i in $Z(S_k)$.

An important role is played by the generating function $T^{(r)}(x)$ for rooted trees whose root vertex has degree r. It follows from Pólya's theorem (see [P37] or Chapter 2 of [HP73]) that

(6.3.23) $$T^{(r)}(x) = xZ(S_r)[s_i \rightarrow T(x^i)].$$

Let $F(x)$ and $f(x)$ count fixed vertices *of degree* r in rooted and unrooted trees. The next theorem summarizes the relations Bailey found for these series.

Theorem 6.3.6. The generating functions for fixed vertices of degree r in rooted and unrooted trees are

(6.3.24) $$F(x) = T^{(r)}(x) + T(x)[O_1(x) - O_1(x^2)]$$

and

(6.3.25) $$f(x) = T^{(r)}(x) - [1 + T(x)]O_1(x^2),$$

where

(6.3.26) $$O_1(x) = F(x) + T^{(r-1)}(x) - T^{(r)}(x).$$

The easiest proof is patterned after the approach used earlier for fixed vertices alone but is a bit more involved (see [Ba82] for details).

For orbit sizes $s \geq 2$, let $O_s(x)$ count vertices of degree r and orbit size s in rooted trees; $o_s(x)$ does the same for unrooted trees. Here are the elegant formulas discovered by Bailey for these series. Note that the degree r is suppressed.

Theorem 6.3.7. The generating functions for vertices of orbit size $s \geq 2$ and degree r for rooted and unrooted trees are

(6.3.27) $$O_s(x) = T(x)\sum_{d|s} d\left[O_{s/d}(x^d) - O_{s/d}(x^{d+1})\right]$$

and

(6.3.28) $$o_s(x) = O_s(x) - [T(x)O_s(x) + O_s(x^2)] + \begin{cases} 0 & s \text{ odd} \\ O_{s/2}(x^2) & s \text{ even.} \end{cases}$$

As before, a combinatorial proof similar to the one we used on (6.3.18) can be used to establish (6.3.27), and the Otter approach can be employed on (6.3.28).

The companion theorem for asymptotic behavior can now be derived. Note that we use the observation of [RoS75]:

$$(6.3.29) \qquad T^{(r)}(x) \sim T(x)\rho Z(S_{r-1})\big[s_i \to T(\rho^i)\big].$$

Theorem 6.3.8. The asymptotic probability of a fixed vertex of degree r in trees of order n is

$$(6.3.30) \qquad \frac{r_n}{nt_n}\big\{\rho Z(S_{r-1})\big[s_i \to T(\rho^i)\big] - O_1(\rho^2)\big\},$$

and for orbit size $s \geq 2$ and degree r it is

$$(6.3.31) \qquad \frac{r_n}{nt_n}\big\{\textstyle\sum k\big[O_{s/k}(\rho^k) - O_{s/k}(\rho^{k+1})\big] - O_s(\rho^2)\big\},$$

where the sum is over all divisors k of s with $k \neq 1$.

Application of the theorem required some nice computing, and the percentages found by Bailey are displayed in Table 6.3.2. There we see, for

TABLE 6.3.2. Percentages of Vertices by Orbit Size and Degree in a Large Unlabeled Random Tree

Orbit Size	Degree 1		Degree 2		Degree 3		Degree 4	
1	17.1497	E + 00	26.3995	E + 00	15.5747	E + 00	68.2144	E − 01
2	14.4257	E + 00	25.7361	E − 01	32.0780	E − 02	37.5820	E − 03
3	66.1506	E − 01	35.8108	E − 02	15.0104	E − 03	59.5924	E − 05
4	34.6092	E − 01	60.1067	E − 03	84.5852	E − 05	11.3350	E − 06
5	12.1206	E − 01	72.2350	E − 04	34.7393	E − 06	15.8061	E − 08
6	61.7470	E − 02	11.9858	E − 04	19.2696	E − 07	29.5426	E − 10
7	19.3361	E − 02	13.1361	E − 05	72.3426	E − 09	37.6809	E − 12
Total	43.6742	E 00	29.3999	E 00	15.9114	E 00	68.5963	E − 01

Orbit Size	Degree 5		Degree 6		Degree 7		Total	
1	25.9836	E − 01	92.5431	E − 02	31.9687	E − 02	69.7889	E 00
2	43.3320	E − 04	49.7178	E − 05	56.9537	E − 06	17.3625	E 00
3	23.2685	E − 06	90.3576	E − 08	35.0237	E − 09	69.8880	E − 01
4	14.9619	E − 08	19.6516	E − 10	25.7674	E − 12	35.2189	E − 01
5	70.6728	E − 11	31.4176	E − 13	13.9396	E − 15	12.1932	E − 01
6	44.6242	E − 13	66.8834	E − 16	87.9146	E − 19	61.8670	E − 02
7	19.2853	E − 15	98.1329	E − 19	49.8375	E − 22	19.3493	E − 02
Total	26.0272	E − 01	92.5930	E − 02	31.9744	E − 02	99.6936	E 00

example, that about 17% of the vertices are fixed and have degree 1. Further-more, the sum of the asymptotic percentages for $r, s \leq 7$ accounted for 99.7% of all the vertices in the random tree.

In the miniseries [BaKP81] and [BaKP83], the techniques sketched here were applied to several families of trees. We calculated tables of asymptotic proportions for vertices of degree r and orbit size s for trees of maximum degree 3 and 4 and also for those whose degrees are all 1 or 3 [called (1, 3)-trees] and 1 or 4. In [PaR-P] we will provide still more information about the nature of the symmetries of a random tree of specified maximum degree. We used an idea involving a two-variable logarithmic generating function that seems to have originated in the work of Etherington [Et38]. To illustrate, let $t(x, y)$ be the generating function in two variables x and y such that the coefficient of $x^n y^m$ is the number of trees T of order n in which m is the logarithm base 2 of the order of the automorphism group of T. Then $t(x, 2)$ counts automorphism in these trees. The 20-step algorithm [HRS75] can be applied, and we find, for example, that the sum of the orders of the groups of all trees with maximum degree 3 is asymptotic to

$$(6.3.32) \qquad\qquad \frac{c}{n^{5/2}\alpha^n},$$

where c is a constant and $\alpha = .384\ldots$ is the radius of convergence of $t(x, 2)$. It follows from Otter's results (see [O48] or [BaKP83]) that the expected group order for these trees increases exponentially.

We conclude by emphasizing that these techniques cannot be applied without some modification to the problem of estimating the group order of any tree of large order n. The reason is that the series for counting group orders does not converge, since the coefficient of x^n is at least $(n - 1)!$.

Exercises 6.3

1. Give a complete justification of (6.3.3), including a proof of Clarke's formula (6.3.4) based on an application of Prüfer's multinomial coefficients (6.1.1).

2. Derive the recurrence relation (6.1.20) for rooted trees from Cayley's relation (6.3.7).

3. Use (6.3.12) to calculate the first 10 coefficients of $D(x)$.

4. Find the numbers of fixed vertices in trees of order $n \leq 10$.

APPENDIXES

*If you can't find it at Ralph's,
you can probably get along without it.*

RALPH OF RALPH'S PRETTY GOOD GROCERY,
LAKE WOBEGON, MINNESOTA

APPENDIX I

NOTATION

Let $\{a_n\}$ and $\{b_n\}$ be sequences of real numbers. Then $a_n \to L$ means $\lim_{n \to \infty} a_n = L$, while $a_n \not\to L$ means the limit is not L.

The big-O and little-o notation is defined as usual:

$$a_n = O(b_n)$$

means that there are constants K and N such that $|a_n| \le Kb_n$ for all $n > N$. Thus $a_n = O(1)$ means $\{a_n\}$ is a bounded sequence. We also have

$$a_n = o(b_n),$$

which means that there is a sequence $\{k_n\}$ of positive terms such that $k_n \to 0$ and a constant N so that $|a_n| \le k_n b_n$ for all $n > N$. For example, if $a_n = o(1)$, then $a_n \to 0$.

If $a_n = b_n(1 + o(1))$, we say that a_n and b_n are *asymptotically equivalent* and we write $a_n \sim b_n$.

We use $\lfloor x \rfloor$ to denote the greatest integer in x, while $\lceil x \rceil$ denotes the least integer that is $\ge x$.

We emphasize that everywhere in the text the letter n stands for the number of vertices in the graphs of our sample spaces and that all limits, big-O and little-o terms are with respect to n unless otherwise specified.

The cardinality of a set S is denoted by $|S|$. A k-set or k-subset has exactly k elements.

We use parentheses throughout for binomial and multinomial coefficients. That is,

$$\binom{n}{k} = \frac{n!}{k!(n-k)!}$$

for nonnegative integers n and k, and similarly

$$\binom{n}{k_1, \ldots, k_m} = \frac{n!}{k_1! \cdots k_n!},$$

where $k_1 + \cdots + k_m = n$.

The end of a proof is signified by an open square $\square$.

APPENDIX II

STIRLING'S FORMULA

The factorial function can be estimated using

$$(2.1) \qquad n! = \sqrt{2\pi n}\left(\frac{n}{e}\right)^n \exp\left\{\frac{\theta}{12n}\right\}$$

where $0 < \theta < 1$ and θ depends on n. Therefore

$$(2.2) \qquad n! = [1 + o(1)]\sqrt{2\pi n}\left(\frac{n}{e}\right)^n.$$

This indispensable weapon of combinatorialists was discovered in the early eighteenth century by James Stirling. Elementary proofs are found in most advanced calculus books. The first step is to obtain upper and lower bounds for $\log n!$ by approximating the integral $\int_1^n \log x\, dx$ with circumscribed and inscribed trapezoids. Then it can be shown that

$$(2.3) \qquad n! = [1 + o(1)]c\sqrt{n}\left(\frac{n}{e}\right)^n$$

for some constant $c > 0$.

Wallis' infinite product for π follows quickly from the integration formulas for $\int_0^{\pi/2} \sin x^n \, dx$ worked out in most calculus texts:

$$(2.4) \qquad\qquad \frac{\pi}{2} = \lim_{n \to \infty} \frac{2^{4n}(n!)^4}{[(2n)!]^2(2n + 1)}.$$

The constant c is now neatly determined by substitution of (2.3) in (2.4). The fine-tuning evident in (2.1) is accomplished by using the usual error term associated with the trapezoidal rule.

See [Fe57] or [Re70] for some of the details and additional references for this important formula.

APPENDIX III

BINOMIAL COEFFICIENTS

Here are a few bounds on the binomial coefficients that we use in the text. As usual, $(n)_k$ denotes the falling factorial:

$$(n)_k = n(n-1) \cdots (n-k+1).$$

First we have

$$(3.1) \qquad \binom{n}{k} = \frac{(n)_k}{k!} = \frac{n^k}{k!} \frac{(n)_k}{n^k} \le \frac{n^k}{k!} \le \left(\frac{en}{k}\right)^k$$

and therefore we concentrate on $(n)_k/n^k$ for better estimates.

Applying the logarithm, exponential and using the Taylor series for $\log(1-x)$, we can find

$$(3.2) \qquad \frac{(n)_k}{n^k} = \exp\left\{-\sum_{i=1}^{k-1} \sum_{m=1}^{\infty} \frac{(i/n)^m}{m}\right\}.$$

Next we use the formulas for the sum of the integers and the sum of the squares of the integers as well as the fact that for m fixed,

$$(3.3) \qquad \sum_{i=1}^{k-1} i^m = O(k^{m+1}),$$

where the big-O term is with respect to k, and we find

$$(3.4) \qquad \frac{(n)_k}{n^k} \sim \exp\left\{-\frac{k^2}{2n} - \frac{k^3}{6n^2}\right\},$$

provided $k = o(n^{3/4})$. Of course, if $k = o(n^{1/2})$, then

$$(3.5) \qquad \frac{(n)_k}{n^k} = 1 + o(1).$$

Furthermore, for any k at all,

$$(3.6) \qquad \frac{(n)_k}{n^k} = O(1)\exp\left\{-\frac{k^2}{2n} - \frac{k^3}{6n^2}\right\}.$$

When working with Model B, one needs estimates of quotients of binomial coefficients. For example, with $0 \le k \le a \le b$,

$$(3.7) \qquad \frac{\binom{a}{k}}{\binom{b}{k}} = \prod_{i=0}^{k-1}\left(1 - \frac{b-a}{b-i}\right) < \exp\left\{-\sum_{i=0}^{k-1}\frac{b-a}{b-i}\right\} < \exp\left\{-\frac{b-a}{b}k\right\}.$$

In fact, if k, a and b are functions of n, it can be shown with a little more effort that

$$(3.8) \qquad \frac{\binom{a}{k}}{\binom{b}{k}} \sim \exp\left\{-\frac{b-a}{b}k\right\},$$

provided

$$k^2\frac{b-a}{b^2} \to 0$$

and

$$k\left(\frac{b-a}{b-k}\right)^2 \to 0.$$

This estimate has the following useful special case for Model B:

$$(3.9) \qquad \frac{\left(\!\!\binom{\binom{n-r}{2}}{q}\!\!\right)}{\left(\!\!\binom{\binom{n}{2}}{q}\!\!\right)} \sim e^{-2rq/n}.$$

provided $q = o(n^{3/2})$ and $r = o(\omega_n)$, where $\omega_n \to \infty$ arbitrarily slowly.

Here is another useful formula for $k \le b \le a$:

$$(3.10) \qquad \frac{\binom{a-k}{b-k}}{\binom{a}{b}} = \frac{(b)_k}{(a)_k} \le \left(\frac{b}{a}\right)^k.$$

If k, b and a are functions of n, then

$$(3.11) \qquad \frac{\binom{a-k}{b-k}}{\binom{a}{b}} \sim \left(\frac{b}{a}\right)^k,$$

provided $k = o(b^{1/2})$ and $k = o(a^{1/2})$. The following application could be used in Model B to deal with the expected number of cycles. Suppose $q \sim cn$ for constant $c > 0$ and $k = o(n^{1/2})$; then

$$(3.12) \qquad \frac{\binom{\binom{n}{2}-k}{q-k}}{\binom{\binom{n}{2}}{q}} \sim \left(\frac{2q}{n^2}\right)^k.$$

APPENDIX IV

BINOMIAL DISTRIBUTION

Following Feller [Fe57], we sometimes use the notation

$$(4.1) \qquad b(k; n, p) = \binom{n}{k} p^k (1 - p)^{n-k},$$

where $0 < p < 1$ and $0 \le k \le n$. We begin with an estimate of the upper and lower tails of the binomial distribution (see [Fe57], p. 140).

Proposition 4.1. For $r \ge pn$,

$$(4.2) \qquad \sum_{k=r}^{n} b(k; n, p) \le b(r; n, p) \frac{(r + 1)(1 - p)}{(r + 1) - p(n + 1)},$$

and for $s \le pn$,

$$(4.3) \qquad \sum_{k=0}^{s} b(k; n, p) \le b(s; n, p) \frac{p(n - s + 1)}{p(n + 1) - s}.$$

Proof. First let $t_k = b(k; n, p)$ and apply the ratio test:

$$(4.4) \qquad \frac{t_{k+1}}{t_k} = \frac{(n - k)p}{(k + 1)(1 - p)} = 1 + \frac{(n + 1)p - (k + 1)}{(k + 1)(1 - p)}.$$

Note that the ratio decreases as k increases. Therefore, for all $k \ge r$,

$$(4.5) \qquad \frac{t_{k+1}}{t_k} \le \frac{(n - r)p}{(r + 1)(1 - p)} = \alpha.$$

Now $\alpha < 1$ for $r \geq pn$, and so we can bound the tail by a convergent geometric series. Just observe that

$$(4.6) \qquad \frac{t_{r+m}}{t_r} = \prod_{k=r}^{r+m-1} \frac{t_{k+1}}{t_k} \leq \alpha^m.$$

Therefore

$$(4.7) \qquad \sum_{m=0}^{n-r} \frac{t_{r+m}}{t_r} < \sum_{m=0}^{n-r} \alpha^m,$$

or

$$(4.8) \qquad \sum_{k=r}^{n} t_k < t_r \frac{1}{1-\alpha},$$

and after substitution for α we have the required bound for the upper tail. The bound for the lower tail is derived by symmetry. That is, apply (4.2) with $r = n - s$ and p and $(1 - p)$ interchanged. $\square$

Next the bounds in this proposition are approximated. We set $\delta = \delta_k = k - pn$ so that $\delta = 0$ when k is equal to the mean, pn, of the binomial distribution. Then Stirling's formula can be used to derive the relation

$$
b(k; n, p) \sim \left\{ 2\pi n \left(p + \frac{\delta}{n} \right) \left[(1 - p) - \frac{\delta}{n} \right] \right\}^{-1/2} \left(1 + \frac{\delta}{pn} \right)^{-(pn+\delta)}
$$
$$(4.9)$$
$$
\times \left(1 - \frac{\delta}{(1-p)n} \right)^{-[(1-p)n-\delta]}.
$$

Taking logs, one can show

$$
\left(1 + \frac{\delta}{pn} \right)^{-(pn+\delta)} \left(1 - \frac{\delta}{(1-p)n} \right)^{-[(1-p)n-\delta]}
$$
$$(4.10)$$
$$
= \left[\exp\left\{ \frac{\delta^2}{pn} \sum_{i=0}^{\infty} (-1)^i \frac{(\delta/pn)^i}{(i+1)(i+2)} \right. \right.
$$
$$
\left. \left. + \frac{\delta^2}{(1-p)n} \sum_{i=0}^{\infty} (-1)^i \frac{[\delta/(1-p)n]^i}{(i+1)(i+2)} \right\} \right]^{-1}.
$$

This last expression is asymptotic to

$$(4.11) \qquad \exp\left\{ - \frac{\delta^2}{2p(1-p)n} \right\}$$

provided $\delta^3/(pn)^2$ and $\delta^3/[(1-p)n]^2$ are both $o(1)$.

As for the other factor,

$$(4.12) \quad \left\{ 2\pi n \left(p + \frac{\delta}{n} \right) \left[(1 - p) - \frac{\delta}{n} \right] \right\}^{-1/2} \sim [2\pi p(1 - p)n]^{-1/2}$$

provided $\delta/(pn)$ and $\delta/[(1 - p)n]$ are both $o(1)$.

Therefore if $0 < p < 1$, p is fixed and $\delta = o(n^{2/3})$, then we have the well-known estimate (see [Fe57], p. 170)

$$(4.13) \quad b(k; n, p) \sim \frac{1}{\sqrt{2\pi np(1 - p)}} \exp\left\{ - \frac{\delta^2}{2np(1 - p)} \right\}.$$

Now suppose $\delta \sim \varepsilon pn$ for some fixed ε with $0 < \varepsilon < 1$. Also assume $pn \to \infty$ but $p^2 n \to 0$. Therefore

$$(4.14) \quad \frac{\delta^2}{(1 - p)n} \sim \varepsilon^2 p^2 n = o(1)$$

and so

$$(4.15) \quad \begin{aligned} &\left(1 + \frac{\delta}{pn} \right)^{-(pn+\delta)} \left(1 - \frac{\delta}{(1 - p)n} \right)^{-[(1-p)n-\delta]} \\ &\sim \exp\left\{ -\varepsilon^2 pn \sum_{i=0}^{\infty} (-1)^i \frac{\varepsilon^i}{(i + 1)(i + 2)} \right\} \\ &\le \exp\left\{ - \frac{\varepsilon^2 pn}{3} \right\}. \end{aligned}$$

If $\delta \sim -\varepsilon pn$, then

$$(4.16) \quad \begin{aligned} &\left(1 + \frac{\delta}{pn} \right)^{-(pn+\delta)} \left(1 - \frac{\delta}{(1 - p)n} \right)^{-[(1-p)n-\delta]} \\ &\sim \exp\left\{ -\varepsilon^2 pn \sum_{i=0}^{\infty} \frac{\varepsilon^i}{(i + 1)(i + 2)} \right\} \\ &\le \exp\left\{ - \frac{\varepsilon^2 pn}{2} \right\}. \end{aligned}$$

Also, if $\delta \sim \pm \varepsilon pn$, then

$$(4.17) \quad \left\{ 2\pi n \left(p + \frac{\delta}{n} \right) \left[(1 - p) - \frac{\delta}{n} \right] \right\}^{-1/2} = O(1)(pn)^{-1/2}.$$

Therefore if $\delta \sim \pm \varepsilon pn$,

(4.18) $$b(k; n, p) = O(1)(pn)^{-1/2}\exp\left\{-\frac{\varepsilon^2 pn}{3}\right\}.$$

This bound will be useful in obtaining a rough estimate of the interval containing the degrees of the vertices of a random graph (see Chapter 5).

When p is fixed, the estimate of (4.13) can be used (see [Fe57], pp. 168–172) to relate the binomial to the normal distribution in the De Moivre–Laplace limit theorem. Here we will just state the special version that approximates the upper tail. There is a slight shift in notation. Above we had $\delta = \delta_k = k - pn$. The variable $x = x_k$ below is related to δ by the equation

(4.19) $$x = \frac{\delta}{\sqrt{p(1-p)n}}.$$

Proposition 4.2. If $0 < p < 1$ is fixed, $n \to \infty$, $x \to \infty$, but $x^3/\sqrt{p(1-p)n} \to 0$, then

(4.20) $$\sum b(k; n, p) = [1 + o(1)]\frac{1}{\sqrt{2\pi}}\frac{e^{-x^2/2}}{x},$$

where the sum is over all k such that

(4.21) $$k \geq \lfloor pn + x\sqrt{p(1-p)n} \rfloor.$$

Some very useful estimates of the tail now follow from this proposition (see [Bo79], p. 134). Take $x = c(\log n)^{1/2}$, where $c = c(n) = O(1)$ and find

(4.22) $$\sum b(k; n, p) = [1 + o(1)](2\pi c^2 \log n)^{-1/2} n^{-c^2/2}$$

where the sum is over all k such that

(4.23) $$k \geq \lfloor pn + c[p(1-p)n \log n]^{1/2} \rfloor.$$

With only a little more effort, the sums of the squares of these terms can be estimated (see also [Bo79]):

(4.24) $$\sum b(k; n, p)^2 = [1 + o(1)][8\pi p(1-p)c^2 n \log n]^{-1/2} n^{-c^2},$$

where the sum is again over all k that satisfy (4.23).

APPENDIX V

PROBABILITY THEORY

Let Ω be a finite set called the *sample space*. An element of Ω is denoted by G (think graph), and its probability is $P(G)$. By definition,

$$(5.1) \qquad P(\Omega) = \sum_{G \in \Omega} P(G) = 1.$$

An *event* $\mathscr{A}$ is just a subset of the sample space Ω, and its probability $P(\mathscr{A})$ is the sum of the probabilities of its elements. The complementary event is $\overline{\mathscr{A}}$.

A *random variable*, $X = X(G)$, is a real-valued function defined on Ω. Events are often described in terms of random variables. For example, let $\mathscr{A}$ consist of all $G \in \Omega$ such that $X(G) \geq k$. Then the probability of this event is denoted by either $P(\mathscr{A})$ or $P(X \geq k)$.

The *expected value* of X is defined by

$$(5.2) \qquad E(X) = \sum_{G \in \Omega} X(G)P(G).$$

Sometimes $E(X)$ is called the *expectation* or *mean* and is also denoted by μ. Note that $E(\ )$ is linear. That is, for any two random variables, X, Y and real numbers a, b:

$$(5.3) \qquad E(aX + bY) = aE(X) + bE(Y).$$

The *variance* of X is defined by

$$(5.4) \qquad V(X) = E\big((X - E(X))^2\big)$$

and since it is nonnegative, it is occasionally denoted by σ^2.

Proposition 5.1

$$V(X) = E(X^2) - E(X)^2.$$

This formula follows immediately from the linearity of expected value.

Proposition 5.2.   (Markov's inequality).   If $X \geq 0$ and $t > 0$,

$$P(X \geq t) \leq \frac{E(X)}{t}.$$

This simple but extremely useful inequality also follows quickly from the definition of $E(X)$. An immediate consequence of it is the next proposition, which we use many times. Usually it is applied to a sequence $\{\Omega_n\}$ of sample spaces where it is understood that for each n there is a random variable X defined on Ω_n. Just set $t = 1$ in Proposition 5.2.

Proposition 5.3.   If X is nonnegative, then

$$P(X \geq 1) \leq E(X),$$

and if X is also integer-valued, then $E(X) \to 0$ implies $P(X = 0) \to 1$.

Obviously, $E(X) \to 0$ implies $P(X \geq 1) \to 0$. In graph theoretic applications, X is usually both nonnegative and integer-valued, so

$$P(X \geq 1) + P(X = 0) = 1.$$

Thus $P(X = 0) \to 1$.

If Markov's inequality is applied to the random variable $Y = [X - E(X)]^2$, we obtain yet another very useful bound.

Proposition 5.4.   (Chebyshev's inequality).   For $t > 0$,

$$P(|X - E(X)| \geq t) \leq \frac{V(X)}{t^2}.$$

Note that if $X(G) = 0$, then G satisfies the condition on the left side of Chebyshev's inequality when $t = |E(X)|$. Therefore Propositions 5.1 and 5.4 can be combined to obtain the next result.

Proposition 5.5.   If $E(X) \neq 0$,

$$P(X = 0) \leq \frac{E(X^2)}{E(X)^2} - 1.$$

Note that if we are dealing with a sequence of sample spaces and associated random variables, then $E(X^2) \sim E(X)^2$ implies $P(X = 0) \to 0$. Thus we have methods from Propositions 5.3 and 5.5 that will show $P(X = 0) \to 1$ or $P(X = 0) \to 0$.

We use the falling factorial notation:

(5.5) $$(X)_0 = X^0 = 1,$$

and for the positive integer r,

(5.6) $$(X)_r = X(X - 1) \cdots (X - r + 1).$$

Then the Stirling numbers $s(r, j)$ of the first kind are defined by

(5.7) $$s(0,0) = 1$$

and

(5.8) $$(X)_r = \sum_{j=1}^{r} s(r, j) X^j, \qquad r > 0,$$

while the Stirling numbers $S(r, j)$ of the second kind are given by

(5.9) $$S(0,0) = 1$$

and

(5.10) $$X^r = \sum_{j=1}^{r} S(r, j)(X)_j, \qquad r > 0.$$

Suppose the random variable is expressed as a sum

(5.11) $$X = X_1 + \cdots + X_{l_n}$$

where for each $i = 1$ to l_n, $X_i(G) = 0$ or 1. Then the "binomial moments" are defined by

(5.12) $$S_0 = 1$$

and for the positive integer r

(5.13) $$S_r = \sum_1 E(X_{l_1} \cdots X_{l_r}),$$

where the sum $\sum_1$ is over all $1 \leq l_1 < \cdots < l_r \leq l_n$, that is, all r-subsets of $\{1, \ldots, l_n\}$.

The rth "factorial moments" are defined by

(5.14) $$E_r(X) = E((X)_r)$$

and these are related to the binomial moments by the next proposition.

Proposition 5.6

$$S_r = \frac{E_r(X)}{r!}$$

To verify Proposition 5.6, start with

(5.15) $$E_r(X) = \sum_{j=r}^{l_n} (j)_r P(X = j).$$

Now divide both sides by $r!$ to obtain

(5.16) $$\frac{E_r(X)}{r!} = \sum_{j=r}^{l_n} \binom{j}{r} P(X = j)$$

$$= \sum_1 P(X_{l_1} \cdots X_{l_r} = 1)$$

$$= S_r.$$

The probabilistic form of the principle of inclusion and exclusion can be stated in terms of the binomial moments.

Proposition 5.7.   (Inclusion and exclusion)

$$P(X = 0) = \sum_{i=0}^{l_n} (-1)^i S_i.$$

We frequently use the following more general result.

Proposition 5.8.   (Ch. Jordan's formula)

$$P(X = k) = \sum_{i=0}^{l_n - k} (-1)^i \binom{i + k}{i} S_{i+k}.$$

In addition to the exact formulas above, we also have the following inequalities.

Proposition 5.9.   (Bonferroni inequalities).   For each k and m,

$$P(X = k) \le \sum_{j=0}^{2m} (-1)^j \binom{k + j}{j} S_{k+j}$$

and

$$P(X = k) \geq \sum_{j=0}^{2m-1} (-1)^j \binom{k+j}{j} S_{k+j}.$$

Note that for $k = 0$,

(5.17) $$\sum_{j=0}^{2m-1} (-1)^j S_j \leq P(X = 0) \leq \sum_{j=0}^{2m} (-1)^j S_j.$$

In dealing with random graphs, we always have a sequence $\{\Omega_n\}$ of sample spaces with random variable X defined on each one. Usually for each n, X is written as a direct sum, so that the binomial moment S_r is also defined for each n. Now suppose that for each $r = 1, 2, \ldots,$

(5.18) $$\lim_{n \to \infty} S_r = \frac{\mu^r}{r!}, \qquad 0 < \mu < \infty.$$

Then the Bonferroni inequalities show that for fixed k and m,

(5.19) $$\lim_{n \to \infty} P(X = k) \leq \sum_{j=0}^{2m} (-1)^j \binom{k+j}{j} \frac{\mu^{k+j}}{(k+j)!}.$$

Therefore

(5.20) $$\lim_{n \to \infty} P(X = k) \leq \frac{\mu^k}{k!} \sum_{j=0}^{2m} \frac{(-\mu)^j}{j!}.$$

for *any m*.
Thus

(5.21) $$\lim_{n \to \infty} P(X = k) \leq \frac{e^{-\mu} \mu^k}{k!}$$

and similarly for the reverse inequality. Then we can conclude

(5.22) $$P(X = k) \to \frac{e^{-\mu} \mu^k}{k!}$$

and we say that X is distributed in the limit according to Poisson's law with mean μ. Stronger statements can be made if $S_r \to \mu^r/r!$ uniformly in r.

Most of the material in this section, as well as proofs we have omitted, can be found in the basic texts of Feller [Fe57], Rényi [Re70] or Riordan [Ri58].

APPENDIX VI

RELATION BETWEEN MODELS A AND B

Bollobás [Bo79] has provided the following results, which show the relationship between Models A and B. A set $\mathscr{A}$ of graphs is called *convex* if $G \in \mathscr{A}$ whenever G_1 and G_2 are in $\mathscr{A}$ and the subgraph relation $G_1 \subseteq G \subseteq G_2$ is satisfied.

We assume that $p = p(n)$, with $0 < p < 1$, is the probability of an edge. Furthermore, we suppose $pn^2 \to \infty$ and $(1 - p)n^2 \to \infty$. These two conditions just guarantee that almost every graph has edges and non-edges.

Theorem 6.1. Suppose $\mathscr{A}$ is a set of graphs of order n with property Q and $\varepsilon > 0$ is fixed. Furthermore, assume that if $q = q(n)$ is any sequence of integers such that

$$(6.1) \qquad (1 - \varepsilon)p\binom{n}{2} < q < (1 + \varepsilon)p\binom{n}{2},$$

then $P(\mathscr{A}) \to 1$ in Model B, that is, almost every graph has property Q. Then also, in Model A, almost every graph has property Q. Now suppose $\mathscr{A}$ is a convex set, and in Model A almost every graph has property Q. Then if we set $q = \left\lfloor p\binom{n}{2} \right\rfloor$, in Model B almost every graph has property Q.

Here is an illustration. Consider the convex set $\mathscr{C}$ of connected graphs. In Chapter 4 we showed that if $p = c(\log n)/n$ with $c > 1$, almost every graph is

143

connected. Just let $q = q(n)$ be a sequence of integers such that

$$(6.2) \qquad\qquad q \sim p\binom{n}{2} \sim c\tfrac{1}{2}n \log n.$$

Then with this sequence and $c > 1$, in Model B almost every graph is connected. In fact, the sharp threshold, $p = (\log n)/n + x/n$, for connectivity can be translated to Model B (see [ErR59]) in the same fashion:

$$(6.3) \qquad\qquad q \sim \binom{n}{2}\left(\frac{\log n}{n} + \frac{x}{n}\right) \sim \frac{1}{2}(n \log n + nx).$$

With this function for q, in Model B, the probability of connectivity in the limit is given by

$$(6.4) \qquad\qquad P(\mathscr{C}) \to e^{-e^{-x}},$$

just as in Model A.

Since Models A and B are so closely related, Model A is often chosen because computations are so much easier.

APPENDIX VII

INCLUSION AND EXCLUSION

Let $\mathcal{U}$ be the universal set of S_0 elements, and suppose that $A_1, \ldots, A_m$ are m subsets of $\mathcal{U}$. The complement of A_i in $\mathcal{U}$ is denoted by $\bar{A}_i$. For $r = 1$ to m, define

$$(7.1) \qquad S_r = \sum |A_{l_1} \cap \cdots \cap A_{l_r}|,$$

where the sum is over all $1 \leq l_1 < \cdots < l_r \leq m$, that is, all r-subsets of $\{1, \ldots, m\}$.

Now for $k = 0$ to m let N_k be the number of elements of $\mathcal{U}$ that belong to exactly k of the sets $A_1, \ldots, A_m$. That is,

$$(7.2) \qquad N_k = \left| \left\{ u \in \mathcal{U} \mid k = \left| \{ j \mid u \in A_j \} \right| \right\} \right|.$$

For $1 \leq r, k \leq m$, the numbers S_r and N_k are very closely related, and this relation is neatly expressed in terms of the polynomials

$$(7.3) \qquad S(x) = \sum_{i=0}^{m} S_i x^i$$

and

$$(7.4) \qquad N(x) = \sum_{i=0}^{m} N_i x^i.$$

Proposition 7.1.

$$N(x + 1) = S(x).$$

Proof. The coefficient of x^k in the polynomial $P(x)$ is denoted by $[x^k]P(x)$. Then

(7.5)
$$[x^k]N(x + 1) = \sum_{i=0}^{m} N_i[x^k](x + 1)^i$$

$$= \sum_{i=0}^{m} N_i[x^k] \sum_{j=0}^{i} \binom{i}{j} x^j$$

$$= \sum_{i=k}^{m} N_i \binom{i}{k}.$$

We must show that the right side of (7.5) is S_k. Consider any element u of $\mathcal{U}$ and suppose u belongs to exactly the t sets $A_{i_1}, \ldots, A_{i_t}$. If $t < k$, then u does not belong to any intersection of k sets and so contributes nothing to S_k. but if $t \geq k$, it will contribute 1 for each collection of k sets selected from the t sets above, for a total of $\binom{t}{k}$. There are exactly N_t elements like u, and so their overall contribution to S_k is $\binom{t}{k} N_t$. On summing this expression for $t = k$ to m, we have

(7.6)
$$S_k = \sum_{t=k}^{m} N_t \binom{t}{k}. \qquad \square$$

All sorts of goodies now fall out of this proposition. First set $x = -1$ in Proposition 7.1 and obtain

(7.7)
$$N_0 = \sum_{i=0}^{m} (-1)^i S_i.$$

This is the usual expression for the number N_0 of elements in $\mathcal{U}$ that belong to none of the sets $A_1, \ldots, A_m$. Now replace x b $x - 1$ in Proposition 7.1, and compare coefficients of x^k. One finds that

(7.8)
$$N_k = \sum_{i=0}^{m-k} (-1)^i \binom{i + k}{i} S_{i+k}.$$

These two formulas should be compared with their probabilistic counterparts, Propositions 5.7 and 5.8 of Appendix V.

It is important to study the behavior of the partial sums of $S(-1)$ to determine the effect of including and excluding successive terms. Therefore we consider

$$(7.9) \qquad \sum_{j=0}^{s} (-1)^j S_j = \sum_{j=0}^{s} (-1)^j \sum_{i=j}^{m} N_i \binom{i}{j},$$

where the right side has been obtained by substitution of (7.6). Now we interchange the order of summation and obtain

$$(7.10) \qquad \sum_{j=0}^{s} (-1)^j S_j = \sum_{i=0}^{m} N_i \sum_{j=0}^{s} (-1)^j \binom{i}{j}.$$

Note that $\binom{i}{j} = 0$ whenever $i < j$, but $\binom{0}{0} = 1$. The following formula is easy to verify:

$$(7.11) \qquad \sum_{j=0}^{s} (-1)^j \binom{i}{j} = \begin{cases} (-1)^s \binom{i-1}{s} & i \geq 1 \\ 1 & i = 0. \end{cases}$$

Now one more substitution gives us

$$(7.12) \qquad \sum_{j=0}^{s} (-1)^j S_j = N_0 + (-1)^s \sum_{i=1}^{m} N_i \binom{i-1}{s}.$$

Since the contribution of the sum on the right side of (7.12) is negative or positive according as s is odd or even, we can write

$$(7.13) \qquad \sum_{j=0}^{2s-1} (-1)^j S_j \leq N_0 \leq \sum_{j=0}^{2s} (-1)^j S_j.$$

The generalization of this entrapment is given by

$$(7.14) \qquad \sum_{j=0}^{2s-1} (-1)^j \binom{k+j}{j} S_{k+j} \leq N_k \leq \sum_{j=0}^{2s} (-1)^j \binom{k+j}{j} S_{k+j}.$$

The corresponding probabilistic formulas are in Proposition 5.9 of Appendix V.

APPENDIX VIII

GRAPH THEORY

LABELED GRAPHS

Let $V = \{v_1, v_2, \ldots, v_n\}$ be a finite set of n elements called *vertices*, and let E be a subset of the collection of $\binom{n}{2}$ unordered pairs of vertices. The elements of E are called *edges*. A *labeled graph* G consists of the ordered pair (V, E) of sets of vertices and edges. The number n of vertices is the *order* of G, while the number q of edges is called the *size*. The number of labeled graphs of order n and size q with vertex set V is denoted by $G_{n,q}$. It equals the number of q-subsets of the collection of $\binom{n}{2}$ unordered pairs of vertices, and hence

$$(8.1) \qquad\qquad G_{n,q} = \left(\binom{\binom{n}{2}}{q} \right).$$

The number G_n of labeled graphs of order n is the total number of these subsets, and therefore

$$(8.2) \qquad\qquad G_n = 2^{\binom{n}{2}}.$$

We always use the letters u, v and w, perhaps with subscripts, to denote vertices. An edge formed by the vertices u and v is denoted by $\{u, v\}$, and u and v are said to be *adjacent*. The number of vertices adjacent to a given vertex v is called the *degree* of v and is denoted by $\deg v$. An *isolated* vertex has degree 0. An *end-vertex* has degree 1. The minimum degree of a graph G is $\delta = \delta(G)$, while the maximum degree is $\Delta = \Delta(G)$.

The next two propositions are observations made by Euler in his famous paper on the solution of the problem of the Königsberg bridges [BiLW76].

Proposition 8.1. The sum of the degrees of all the vertices of a graph is twice the number of edges.

149

Proposition 8.2.   In any graph the number of vertices of odd degree is even.

Occasionally the letter e, perhaps with a subscript, is used to denote an edge. If e is the edge that consists of the two vertices u and v, then u and e are said to be *incident* with each other. Likewise v is incident with e.

ISOMORPHISM

Two graphs G_1 and G_2 are called *isomorphic* and we write $G_1 \cong G_2$ if there is a one-to-one function φ from the vertex set of G_1 onto the vertex set of G_2 such that for any two vertices u and v of G_1 we have u and v adjacent in G_1 if and only if $\varphi(u)$ and $\varphi(v)$ are adjacent in G_2. The function φ is called an *isomorphism* from G_1 onto G_2.

Since isomorphism is an equivalence relation, the $2^{\binom{n}{2}}$ labeled graphs of order n with vertex set V are partitioned into isomorphism classes called *unlabeled graphs*. Let g_n be the number of these, while $g_{n,q}$ stands for the number of order n and size q. The diagrams for all the unlabeled graphs of order 4 are shown in Figure 1.1.1.

An isomorphism from a graph G onto itself is called an *automorphism*. The set of all automorphisms of a graph is a (finite) permutation group denoted by $\Gamma(G)$ and called the *automorphism group of G*. For example, consider the labeled graph of order 4 and size 5 in which the vertices of degree 3 are v_1 and v_3 but the vertices of degree 2 are v_1 and v_4. Using the usual disjoint cycle notation for permutations, we find four automorphisms:

$$(v_1)(v_2)(v_3)(v_4), \qquad (v_1)(v_3)(v_2 v_4),$$
$$(v_1 v_3)(v_2)(v_4), \qquad (v_1 v_3)(v_2 v_4).$$

It is sometimes convenient to alter the notation slightly by dropping the letter v. Thus we think of the integers $1, \ldots, n$ as the labels of the vertices of G and an automorphism γ as a permutation of these labels which results in a graph γG identical to G. The automorphisms are often called *symmetries*, and their number is denoted by $|\Gamma(G)|$.

TYPES OF GRAPHS

The vertex set of graphs G is sometimes denoted by $V(G)$ and the edge set by $E(G)$. For example, the *complement* of G, denoted $\overline{G}$, has the same vertex set as G. That is, $V(G) = V(\overline{G})$. But two vertices of $\overline{G}$ are adjacent in $\overline{G}$ if and only if they are *not* adjacent in G. Hence $E(\overline{G})$ consists of all unordered pairs of vertices not in $E(G)$. If G and $\overline{G}$ are isomorphic, G is called *self-complementary*.

A *subgraph* H of G is itself a graph whose vertex set $V(H)$ is a subset of $V(G)$, and $E(H)$ is a subset of $E(G)$. If every pair of vertices u and v of H are adjacent in H whenever they are adjacent in G, then H is an *induced* subgraph of G. A subgraph H of G is said to *span* G if it has the same vertex set, that is, $V(H) = V(G)$.

The *complete graph* of order n, denoted K_n, has all $\binom{n}{2}$ possible edges.

In a *bipartite graph* there are two kinds of vertices, namely green or white, and every edge has one green vertex and one white vertex. If there are m green vertices and n white and all mn possible edges are present, we have a *complete bipartite graph*, denoted by $K_{m,n}$.

In a *rooted graph*, one of the vertices, called the *root*, is distinguished. Two rooted graphs are isomorphic if there is an isomorphism from one to the other that preserves not only edges and labels (if present) but also the roots. From time to time it is also handy to consider graphs rooted at an edge.

If every vertex of G has exactly the same degree, say r, G is *regular of degree r* or *r-regular*. A *cubic* graph is regular of degree 3.

CONNECTIVITY FOR GRAPHS

A *walk* in a graph G is a sequence of vertices $w_1, w_2, \ldots, w_m$ such that w_i is adjacent to w_{i+1} for $i = 1$ to $m - 1$. A walk describes the route of a particle that travels from vertex to adjacent vertex along edges in any direction. Vertices and edges may be repeated. A *trail* is a walk in which there are no repeated edges, and a *path* has no repeated vertices (and hence is a trail). A walk (trail or path) beginning at u and ending at v is called a $u - v$ walk (trail or path).

The *length* of a path is the number of edges used. The *distance* between a pair of vertices u and v is the length of a shortest $u - v$ path. The *diameter* of a graph is the maximum of the distances between all pairs of vertices.

A *cycle* is a walk with at least three different vertices that has no repeated vertices except the first and last. Usually there is no confusion if we also refer to the following subgraph H of G as a cycle, where $V(H) = \{w_1, w_2, \ldots, w_m\}$, $E(H) = \{\{w_1, w_2\}, \{w_2, w_3\}, \ldots, \{w_m, w_1\}\}$ and $m \geq 3$. A *unicyclic* graph has exactly one cyclic subgraph. An *acyclic* graph has no cycles. A graph with a spanning cycle is *hamiltonian*.

When a vertex v is *deleted* from the graph G, the graph $G - v$ obtained has as its vertex set all the vertices of G except for v, and its edges are all the edges of G except for those incident with v. If the edge e of a graph G is deleted, the resulting graph, denoted $G - e$, has the same vertex set as G and the same edge set except for e.

A graph is *connected* if there is a path joining every pair of vertices. A *disconnected* graph is not connected. A *component* is a maximal connected subgraph. An edge whose deletion increases the number of components is called a *bridge*. The *connectivity* $\kappa = \kappa(G)$ is the minimum number of vertices whose deletion from G results in a disconnected graph or the trivial graph K_1. The *edge-connectivity* $\lambda = \lambda(G)$ is the minimum number of edges whose deletion results in a disconnected or trivial graph. The following simple inequality relates the connectivity to the minimum degree.

Proposition 8.3

$$\kappa(G) \le \lambda(G) \le \delta(G).$$

The proof follows directly from the definitions.

The following theorem of Menger [Me27] was the first important result found that leads to a deeper understanding of connectivity. We say that two $u - v$ paths are *internally disjoint* if they have no common vertices but u and v. A set S of vertices *separates* u and v if u and v belong to different components of the graph obtained by deleting the set S.

Theorem 8.1. If u and v are different, nonadjacent vertices of G, then the maximum number of internally disjoint $u - v$ paths equals the minimum number of vertices whose deletion separates u and v.

For each integer $m \ge 1$, a graph G is *m-connected* if $\kappa(G) \ge m$. Hence at least m vertices must be deleted to disconnect G or reduce it to K_1. The next result, found by Whitney [Wh32], follows from Menger's theorem.

Corollary 8.1. A graph is *m-connected* if and only if there are at least m internally disjoint paths joining any two vertices.

Dirac [Di60] found some useful properties of *m-connected* graphs that follow easily from Whitney's result.

Corollary 8.2. If G is *m-connected*, there are m internally disjoint paths from any vertex to any m other vertices.

DIGRAPHS

A *labeled digraph* D consists of the set $V = \{v_1, \ldots, v_n\}$ of n vertices and a subset E of the collection of $n(n-1)$ ordered pairs of different elements of V. The elements of E are called *arcs*. In the diagram of a digraph it is customary to draw an arrow from vertex u to vertex v to depict the arc (u, v). If (u, v) is an arc of D, we say that u is *adjacent to* v and v is *adjacent from* u. Occasionally an arc (v, v) from a vertex to itself is permitted. These are called *loops*. The number of vertices adjacent to a given vertex v is called the *indegree* of v and is denoted by *indeg* v. The number of vertices adjacent from v is the *outdegree* of v, denoted outdeg v. Many definitions for digraphs are identical to the corresponding definitions for graphs. We provide only the basic necessities.

CONNECTIVITY FOR DIGRAPHS

A *walk* in a digraph D is a sequence of vertices $w_1, w_2, \ldots, w_m$ such that w_i is adjacent to w_{i+1} for $i = 1$ to $m - 1$. A *trail* is a walk with no repeated arcs, and a *path* has no repeated vertices. The *length* of a walk is the number of arcs in it. A *cycle* is a walk with at least two different vertices that has no repeated vertices except the first and last.

A *semiwalk* is a sequence of vertices $w_1, w_2, \ldots, w_m$ such that w_i is adjacent to *or* from w_{i+1} for each $i = 1$ to m. Thus a particle on a semiwalk in a digraph may occasionally move the wrong way on a one-way street. The notion of a *semitrail* or *semipath* is defined as expected.

A digraph is *weakly connected*, or *weak*, if for any two vertices there is a semipath from one to the other. A digraph is *unilaterally connected*, or *unilateral*, if for any two vertices there is a path from one to the other. Finally, a digraph is *strongly connected*, or *strong*, if for any two vertices u and v there is a path from u to v and also one from v to u.

A *weak component* is a maximal weakly connected subgraph. Unilateral and strong components are similarly described.

A vertex v is *reachable* from u if there is a $u - v$ path. A *source* in a digraph is a vertex from which every other vertex is reachable.

A digraph is *hamiltonian* if it has a spanning cycle.

A *tournament* is a digraph D such that for every two vertices u and v, exactly one of the arcs (u, v) or (v, u) is present in $E(D)$. Think of the vertices as players in a round-robin tournament. If the arc (u, v) is present, player u beats or dominates player v. The following result of Camion [Ca59] plays an important role in determining the proportion of digraphs that are hamiltonian (see exercise 2.3.4).

Theorem 8.   Every strong tournament is also hamiltonian.

TREES

A *tree* is a connected, acyclic graph. Hence the vertices at the ends of any maximal path must both have degree 1, unless the tree is trivial. This observation sets up simple induction proofs for trees by deletion of vertices of degree 1 and leads to many characterizations of trees. Here are two of the most useful descriptions.

Proposition 8.4.   The following statements about the graph G of order n and size q are equivalent:

(i) G is a tree
(ii) G is connected and $n - 1 = q$
(iii) G is acyclic and $n - 1 = q$.

The *eccentricity* of a vertex v of any graph is the distance to a vertex farthest from v. The *center* of the graph consists of all the vertices of minimum eccentricity. Jordan found the center of a tree to be particularly simple (see page 47 of [BiLW76]).

Proposition 8.5.   The center of a tree consists of exactly one vertex or two adjacent vertices.

The proof is made by noticing that the central vertices of a tree are invariant after simultaneous deletion of all vertices of degree 1.

Given a tree T and vertex v, each component of $T - v$ can be regarded as a tree rooted at the vertex adjacent to v in T. These rooted components are called *branches*, and the *weight* of v is the order of the largest branch. The *centroid* of T consists of the vertices of minimum weight. The nature of the centroid was also discovered by Jordan.

Proposition 8.6.   The centroid of a tree consists of exactly one vertex or two adjacent vertices.

To carry out the proof one just shows by contradiction that any two centroidal vertices must be adjacent.

An acyclic graph is called a *forest* because its components must all be trees. Every forest of order n and size q has exactly the same number of components, namely, $n - q$.

CHROMATIC NUMBER

A subset of the vertex set of a graph G is *independent* if the subgraph induced by it has no edges. The *independence number* of G, denoted by $\beta = \beta(G)$, is the maximum order of the independent sets of vertices of G. Suppose the vertex set of G is partitioned into t independent sets. The minimum value of t for which such a partition exists is called the *chromatic number* of G, and it is denoted by $\chi = \chi(G)$. Thus the chromatic number is the smallest number of colors that can be assigned to the vertices of G so that adjacent vertices have different colors. A convenient lower bound for χ can be expressed in terms of the independence number.

Proposition 8.7. For any graph G of order n,

$$\chi(G) \geq \frac{n}{\beta(G)}.$$

Proof. Suppose the vertex set of G is partitioned into t independent sets:

$$V(G) = V_1 \cup \cdots \cup V_t.$$

Then

$$n = \sum_{i=1}^{t} |V_i| \leq t\beta.$$

The proof is finished by having χ for t. □

If we set out to color the vertices of any graph so that adjacent vertices have different colors, even if the task is carried out in the dumbest possible way, $\Delta(G) + 1$ crayons are sufficient to do the job. Brooks [Br41] made an important, nontrivial improvement on this condition.

Theorem 8.3. If G is a complete graph or an odd cycle, then $\chi(G) = \Delta(G) + 1$. Otherwise, $\chi(G) \leq \Delta(G)$.

The *edge-chromatic number* of G, denoted $\chi_1 = \chi_1(G)$, is the minimum number of colors that can be assigned to the edges of G so that no vertex is incident with two edges of the same color. Since every edge incident with a given vertex must have a different color, we have $\chi_1(G) \geq \Delta(G)$. Furthermore a single edge $\{u, v\}$ has at most $2[\Delta(G) - 1]$ other edges incident with u and v, and so $2\Delta(G) - 1$ colors are sufficient for the proper coloring of any edge. Hence we also have an upper bound $\chi_1(G) \leq 2\Delta(G) - 1$. Vizing [Vi64], however, made the surprising discovery that $\chi_1(G)$ has one of just two possible values.

Theorem 8.4. The edge-chromatic number $\chi_1(G)$ is equal to either $\Delta(G)$ or $\Delta(G) + 1$.

We need one more nice result of Vizing [Vi65] that provides a necessary condition for a graph to require the extra color.

Theorem 8.5. If G has $\chi_1(G) = \Delta(G) + 1$, then G has at least three vertices of maximum degree $\Delta(G)$.

Hence a graph with a unique vertex of maximum degree must have $\chi_1(G) = \Delta(G)$.

PLANARITY

A graph G is *planar* if it can be represented in the following way. The vertices of G are points in the plane, and each edge of G is a simple Jordan curve that joins the points corresponding to the vertices of the edge. Furthermore these Jordan curves are mutually disjoint except for the points corresponding to vertices. Thus a graph is planar if its diagram can be drawn so that none of the lines that correspond to edges cross each other except at vertices. A *plane* graph G has already been represented in the plane in this particular way. The set S of points in the plane that do not belong to the representation consists of connected regions called faces. The following special case of Euler's famous formula [BiLW76] relates the number of vertices, edges and faces of a plane graph.

Theorem 8.6. Let G be a connected, plane graph of order n and size q with r faces. Then

$$n - q + r = 2.$$

The proof can be made by deleting an edge of G and inducting on the size.

A plane graph has the maximum number of edges when each face is bounded by three edges. In this case $3r = 2q$, and on substitution in Euler's formula, we have $q = 3n - 6$.

Corollary 8.3.   For any planar graph of order $n \geq 3$ and size q,

$$q \leq 3n - 6.$$

Hence the complete graph K_5 is not planar.

If every face of a plane graph is bounded by four edges, then $4r = 2q$ and hence $q = 2n - 4$.

Corollary 8.4.   For any planar graph of order $n \geq 3$ and size q with no triangles,

$$q \leq 2n - 4.$$

Therefore the bipartite graph $K_{3,3}$ is not planar.

An edge $\{u, v\}$ of a graph is *subdivided* if a new vertex w is introduced and the edge is replaced by two edges $\{u, w\}$ and $\{w, v\}$. Two graphs are *homeomorphic* if each can be obtained by a sequence of subdivisions of the edges of the same graph. From our remarks above it follows that a graph cannot be planar if it contains a subgraph homeomorphic to K_5 or $K_{3,3}$. Kuratowski made the amazing discovery that a graph with no such subgraphs is planar! A full account of the remarkable story behind this important theorem can be found in [BiLW76].

FACTORS

An *r-factor* of a graph is a spanning subgraph that is regular of degree r. Tutte [T47] characterized graphs with a 1-factor.

Theorem 8.7.   A graph of order n has a 1-factor if and only if n is even and for every subset S of the vertex set the number of components of $G - S$ that have odd order is at most $|S|$.

The necessity of this condition is easy to see, but the sufficiency is not at all evident. There is also a very useful application to regular graphs first found for cubics by Petersen [BiLW76].

Corollary 8.5.   Every regular graph of degree $r \geq 3$ with edge-connectivity $\lambda \geq r - 1$ has a 1-factor.

RECONSTRUCTION

The reconstruction conjecture of Ulam is now probably the most famous open problem in graph theory. We will just state the most important versions of the conjecture and refer the reader to the survey of Bondy and Hemminger [BoH77] for more details and appropriate additional references. A graph G with vertex set $V(G) = \{v_1, \ldots, v_n\}$ is *vertex-reconstructible* if any other graph H with $V(H) = \{u_1, \ldots, u_n\}$ is isomorphic to G whenever the vertex-deleted subgraphs $G - v_i$ and $H - u_i$ are isomorphic for $i = 1$ to n.

Note that neither graph of order 2 is vertex-reconstructible because each has the same vertex-deleted subgraphs.

Conjecture 1. (Ulam). Every graph of order $n \geq 3$ is vertex-reconstructible.

Edge-reconstructibility is defined in a similar manner. The two graphs of order 4 and size 2 are seen to have the same edge-deleted subgraphs and hence are not edge-reconstructible. There are also a couple of graphs of order 4 and size 3 that are not edge-reconstructible.

Conjecture 2. (Harary). Every graph of size $q \geq 4$ is edge-reconstructible.

Many special classes of graphs are reconstructible. The following examples are important and easy to determine.

Proposition 8.8. (Harary). All disconnected graphs of order $n \geq 3$ are vertex-reconstructible.

Proposition 8.9. All disconnected graphs of size $q \geq 4$ with at least two nontrivial components are edge-reconstructible.

A subgraph obtained by deleting k vertices is called a *k-vertex-deleted* subgraph. The following result of Chinn [Ch71] may be elementary, but it is crucial in showing that most graphs are vertex-reconstructible (see Chapter 5.4).

Proposition 8.10. If the two-vertex-deleted subgraphs of G are mutually nonisomorphic, then G is vertex-reconstructible.

Proof. First we note that all the one-vertex-deleted subgraphs of G are mutually nonisomorphic. For suppose that we have $G - u \cong G - v$ for different vertices u and v. Let w_1 and w_2 be two vertices of $G - u$ with w_1' and w_2' their corresponding vertices in $G - v$ under this isomorphism. Therefore for $i = 1$ and 2, $G - \{u, w_i\} \cong G - \{v, w_i'\}$. Since two-vertex-deleted subgraphs are nonisomorphic, we must have $u = w_1'$ and $v = w_i$ for $i = 1$ and 2 and hence $w_1 = w_2$, a contradiction.

Thus all the one-vertex-deleted subgraphs of G are different. Consider any two of these, and choose the notation so that one is $G - v_a$ and the other is $G - v_b$. We have only to determine whether or not v_a and v_b are adjacent in G. To do this we call on a third one-vertex-deleted subgraph, say $G - v_i$, to serve as a witness. There exist vertices w_1 in $G - v_a$ and w_1' in $G - v_i$ such that $G - \{v_a, w_1\} \cong G - \{v_i, w_1'\}$. For example, $w_1 = v_i$ and $w_1' = v_a$ will do it. But by hypothesis, they are the *only* vertices that will work. Hence w_1' in $G - v_i$ must be v_a. Similarly v_b can be found in $G - v_i$, and then the question of their adjacency can be settled by $G - v_i$. □

P. K. Stockmeyer [St77] is responsible for the most shocking development in this area. He used a computer to search for counterexamples of order 8 and found the clue that led to an infinite family of nonreconstructible tournaments.

Theorem 8.8. For each $n = 2^m + 1$ or $2^m + 2$, there is a pair of nonisomorphic tournaments of order n that have exactly the same collections of one-vertex-deleted subtournaments.

Hence the vertex-reconstruction conjecture for digraphs is false!

HAMILTONIAN GRAPHS

How can one tell that a given graph is hamiltonian? How does one find a spanning cycle in a hamiltonian graph? These are fundamental problems of graph theory that have generated much research. Some nice answers have been found for the first question, and we will discuss a few of these. But no efficient algorithm has been devised that will produce spanning cycles.

There are quite a few theorems that will identify hamiltonian graphs provided the degrees of the vertices are fairly high. The first of these was discovered by Dirac [Di52].

Theorem 8.9. If G is a graph of order $n \geq 3$ and minimum degree $\delta \geq n/2$, then G is hamiltonian.

Next came Ore's improvement [Or60] of Dirac's theorem.

Theorem 8.10. If G has order $n \geq 3$ and every pair of nonadjacent vertices u and v satisfy

$$\deg u + \deg v \geq n,$$

then G is hamiltonian.

Proof. Fix n, and suppose G is a graph of smallest size for which the theorem fails. Pick any pair u, v of nonadjacent vertices and form the new graph G_1 by adding the edge $\{u, v\}$ to G. That is, $V(G_1) = V(G)$ and $E(G_1) = E(G) \cup \{\{u, v\}\}$. Since G_1 also satisfies the hypothesis of the theorem and is larger in size than G, it must be hamiltonian. Therefore in G there must be a $u - v$ spanning path, say

$$u = w_1, w_2, \ldots, w_n = v.$$

If u is adjacent to w_i for some $i = 2$ to $n - 1$, then v is forbidden to be adjacent to the preceding vertex w_{i-1} on the path. Otherwise G contains the hamiltonian cycle

$$u = w_1, w_i, w_{i+1}, \ldots, w_{n-1}, v, w_{i-1}, \ldots, w_1 = u.$$

Thus there are $\deg u$ forbidden vertices and $\deg v$ adjacent vertices for v. Since there are only $n - 1$ other vertices, $\deg u + \deg v$ must be at most $n - 1$, a contradiction. $\square$

In Pósa's theorem [Po62], which follows, it is assumed that G is a graph of order $n \geq 3$ with vertex set $V = \{v_1, \ldots, v_n\}$, and the names of the vertices have been arranged so that

$$\deg v_1 \leq \deg v_2 \leq \cdots \leq \deg v_n.$$

It is convenient to let $d_i = \deg v_i$ for each $i = 1$ to n, and we call $d_1 \leq d_2 \leq \cdots \leq d_n$ the *degree sequence* of G.

Theorem 8.11. Let G be a graph of order $n \geq 3$ with degree sequence $d_1 \leq d_2 \leq \cdots \leq d_n$ that satisfies the following conditions:

(i) If k is an integer less than $(n - 1)/2$, then $k < d_k$.
(ii) If n is odd and $k = (n + 1)/2$, then $k \leq d_k$.

Then G is hamiltonian.

There are more nice results of this nature by Bondy [Bo69], Chvátal [Ch72] and Las Vergnas [LV76]. Each one relaxes the degree constraints a bit further and is proved by souped-up applications of the pigeonhole principle.

The next criterion of Chvátal and Erdös [ChE72]* is completely different.

Theorem 8.12. If G is a graph of order $n \geq 3$ with connectivity at least as large as the independence number, that is, $\kappa(G) \geq \beta(G)$, then G is hamiltonian.

Proof. If $\beta(G) = 1$, then G must be the complete graph K_n and hence is hamiltonian. Next suppose $\beta(G) \geq 2$ and let $m = \kappa(G) \geq \beta(G) \geq 2$. Now we want to show that G must have a cycle of order at least m. To see this, let P be any maximal path in G and suppose the sequence of vertices in the path is $w_0, w_1, w_2, \ldots, w_r$. Since P is maximal, each vertex adjacent to w_0 must belong to P. From Proposition 8.3 we know that $\delta(G) \geq m$ and therefore w_0 must be adjacent to w_k for some $k \geq m$. This edge $\{w_0, w_k\}$ and the part of the P from w_0 to w_k form a cycle of order at least $m + 1$.

Now let C be a largest cycle of G, and suppose there is a vertex w that does not belong to C. By Dirac's theorem (Corollary 8.2) there are m internally disjoint paths from w to any m vertices of C. Therefore there are m such paths that share with C only their terminal vertices, say $w_1, \ldots, w_m$. No two of these vertices w_i and w_j can be adjacent in C, otherwise the edge $\{w_i, w_j\}$ could be replaced in C by the paths from w_i to w and from w to w_j, thereby constructing a larger cycle than C. Therefore these m vertices of C must be separated by m more vertices of C, say $u_1, u_2, \ldots, u_m$. Now we orient the cycle and choose the notation so that for $i = 1$ to m, vertex u_i is the immediate successor of w_i on C. Note that none of the vertices u_i can be adjacent to w, otherwise a larger cycle could be formed by replacing the edge $\{w_i, u_i\}$ of C with the path from w_i to w and the edge $\{w, u_i\}$.

Consider the set of vertices $S = \{w, u_1, \ldots, u_m\}$. Since the order of S is $m + 1$, it is strictly greater than $\beta(G)$ and so S is not an independent set. But since w is not adjacent to any other vertex u_i of S, we must have u_i and u_j adjacent for some i and j. Again a larger cycle can be constructed from C by deleting the two edges $\{w_i, u_i\}$ and $\{w_j, u_j\}$ and adding the two internally disjoint paths from w_i to w and from w to w_j. Hence a largest cycle in G must be a spanning cycle. $\quad\square$

Lately there have been a number of papers dealing with hamiltonian properties for graphs that do not contain any induced subgraph isomorphic to

---

*This note was written in Professor Richard K. Guy's car on the way from Pullman to Spokane, Washington. The authors have expressed their gratitude to Mrs. Guy for smooth driving, and Louise is honored further in this work by being the only Guy in the list of random graph theorists.

$K_{1,3}$ [MaS84]. We need one more definition. A graph is *locally connected* if for each vertex v of degree at least 2 the subgraph induced by the vertices adjacent to v is connected. Note that a disconnected graph can be locally connected. Here is the condition found by Oberly and Sumner [ObS79].

Theorem 8.13. If G is a graph of order $n \geq 3$ that is connected and locally connected and has no induced subgraph isomorphic to $K_{1,3}$, then G is hamiltonian.

There are other conditions that could be mentioned, but at this point we have enough material for analysis.

MATRICES

The *adjacency matrix* of a graph G of order n is an $n \times n$ matrix $A = (a_{ij})$ in which $a_{ij} = 1$ or 0 according as $\{v_i, v_j\}$ is or is not an edge of G. Thus A is a 0, 1 symmetric matrix with zero diagonal, and the arrangement of 0's and 1's depends on the labeling of the vertices of G. The *characteristic polynomial of G*, denoted $\varphi(G)$, is the characteristic polynomial of the adjacency matrix of G, that is,

$$\varphi(G) = \det(xI - A).$$

Two graphs are *cospectral* if they share the same characteristic polynomial. It used to be conjectured that graphs with cospectral mates were the exception rather than the rule. Schwenk [Sc73] showed that this conjecture was as wrong as possible for trees.

Theorem 8.14. The proportion of trees of order n with cospectral mates has limit 1 as $n \to \infty$; that is, almost all trees have cospectral mates.

Schwenk used enumeration techniques and asymptotic analysis as described in [HRS75] to determine that any specified rooted tree occurs as a branch in almost all trees. To complete the proof he showed how to alter a small branch slightly without changing the spectrum. Naturally the details that back up this clever scheme are quite involved. Schwenk conjectures that almost all graphs are cospectral.

BIBLIOGRAPHY

*If it weren't for Philo T. Farnsworth, inventor of
television, we'd still be eating frozen radio dinners.*

JOHNNY CARSON

This bibliography of about 200 titles owes much to the work of Karoński [Ka82] and Bollobäs [Bo81a]. But it has the journal abbreviations currently employed by *Mathematical Reviews* and the format suggested by the *Journal of Graph Theory* that eliminates superfluous punctuation. The coding scheme that combines the initial letters of an author's name with the year of publication does away with referral errors and allows additions or deletions to be made easily on our word processor.

In addition to the references actually used in the text, quite a few are included here that were erroneous, incomplete or missing from [Ka82] and [Bo81a]. We also pulled many items from MATH FILE and added dozens that have just appeared. Consequently all of the principal references of the field are listed. The scope of a proper bibliography for random graphs would depend much on personal preference, but a rather complete compilation could be based on the references in this text as well as [Ka82], [Bo81a] and the recent survey by Grimmett [Gr83].

[AjKRS82] M. Ajtai, J. Komlós, V. Rödl and E. Semerédi, On coverings of random graphs, *Comment. Math. Univ. Carolin.* **23** (1982) 193–198.

[AjKS79] M. Ajtai, J. Komlós and E. Szemerédi, Topological complete subgraphs in random graphs, *Studia Sci. Math. Hungar.* **14** (1979) 293–297.

163

164 BIBILOGRAPHY

[AjKS81a] M. Ajtai, J. Komlós and E. Szemerédi, A dense infinite Sidon sequence, *Europ. J. Combinatorics* **2** (1981) 1–11.

[AjKS81b] M. Ajtai, J. Komlós and E. Szemerédi, The longest path in a random graph, *Combinatorica* **1** (1981) 1–12.

[AjKS82] M. Ajtai, J. Komlós and E. Szemerédi, Largest random component of a k-cube, *Combinatorica* **2** (1982) 1–7.

[BaES80] L. Babai, P. Erdös and S. M. Selkow, Random graph isomorphism, *SIAM J. Comput.* **9** (1980) 628–635.

[BaK79] L. Babai and L. Kučera, Canonical labelling of graphs in linear average time, *20th Annual IEEE Symp. on Foundations of Comp. Sci.* (Puerto Rico) IEEE, New York (1979) 39–46.

[Ba82] C. K. Bailey, Distribution of points by degree and orbit size in a large random tree, *J. Graph Theory* **6** (1982) 283–293.

[Bar82] A. D. Barbour, Poisson convergence and random graphs, *Math. Proc. Cambridge Philos. Soc.* **92** (1982) 349–360.

[BaKP81] C. K. Bailey, J. W. Kennedy and E. M. Palmer, Points by degree and orbit size in chemical trees I, *The Theory and Applications of Graphs* (G. Chartrand et al., eds.) Wiley, New York (1981) 27–43.

[BaKP83] C. K. Bailey, J. W. Kennedy and E. M. Palmer, Points by degree and orbit size in chemical trees II, *Discrete Appl. Math.* **5** (1983) 157–164.

[BeBK72] A. Békéssy, P. Békéssy and J. Komlós, Asymptotic enumeration of regular matrices, *Studia Sci. Math. Hungar.* **7** (1972) 343–353.

[BeC78] E. A. Bender and E. R. Canfield, The asymptotic number of labeled graphs with given degree sequence, *J. Combin. Theory Ser. A* **24** (1978) 296–307.

[BiLW76] N. L. Biggs, E. K. Lloyd and R. J. Wilson, *Graph Theory 1736–1936*, Clarendon, Oxford (1976).

[BlH79] A. Blass and F. Harary, Properties of almost all graphs and complexes, *J. Graph Theory* **3** (1979) 225–240.

[Bo79] B. Bollobás, *Graph Theory*, Springer, New York (1979).

[Bo80] B. Bollobás, A probabilistic proof of an asymptotic formula for the number of labelled regular graphs, *Europ. J. Combinatorics* **1** (1980) 311–316.

[Bo81a] B. Bollobás, Random Graphs, *Combinatorics, Proc. 8th British Comb. Conf. 1981* (H. N. V. Temperley, ed.) Cambridge (1981) 80–102.

[Bo81b] B. Bollobás, Degree sequences of random graphs, *Discrete Math.* **53** (1981) 1–19.

[Bo81c] B. Bollobás, The diameter of random graphs, *Trans. Amer. Math. Soc.*, **267** (1981) 41–52.

[Bo81d] B. Bollobás, Threshold functions for small subgraphs, *Math. Proc. Cambridge Philos. Soc.* **90** (1981) 197–206.

[Bo81e] B. Bollobás, Counting coloured graphs of high connectivity, *Canad. J. Math.* **33** (1981) 476–484.

[Bo82a] B. Bollobás, Vertices of given degree in a random graph, *J. Graph Theory* **6** (1982) 147–155.

[Bo82b] B. Bollobás, Long paths in sparse random graphs, *Combinatorica* **2** (1982) 223–228.

[Bo82c] B. Bollobás, The asymptotic number of unlabelled regular graphs, *J. London Math. Soc.* (2) **26** (1982) 201–206.

[Bo83] B. Bollobás, Almost all regular graphs are hamiltonian, *Europ. J. Combinatorics* **4** (1983) 97–106.

[Bo-P] B. Bollobás, *Random Graphs*, in preparation.

[BoC81]   B. Bollobás and P. A. Catlin, Topological cliques of random graphs, *J. Combin. Theory Ser. B* **30** (1981) 224–227.

[BoCE80]   B. Bollobás, P. A. Catlin and P. Erdös, Hadwiger's conjecture is true for almost every graph, *Europ. J. Combinatorics* **1** (1980) 195–199.

[BoE76]   B. Bollobás and P. Erdös, Cliques in random graphs, *Math. Proc. Cambridge Philos. Soc.* **80** (1976) 419–427.

[BoK84]   B. Bollobás and V. L. Klee, Diameters of random bipartite graphs, *Combinatorica* **4** (1984) 7–19.

[BoP77]   B. Bollobás and E. M. Palmer, Almost all simplicial complexes are pure, *Utilitas Math.* **12** (1977) 101–111.

[BoT81]   B. Bollobás and A. G. Thomason, Graphs which contain all small graphs, *Europ. J. Combinatorics* **2** (1981) 13–15.

[BoV82]   B. Bollobás and W. F. de la Vega, The diameter of random regular graphs, *Combinatorica* **2** (1982) 125–134.

[Bo69]   J. A. Bondy, Properties of graphs with constraints on degrees, *Studia Sci. Math. Hungar.* **4** (1969) 473–475.

[BoH77]   J. A. Bondy and R. L. Hemminger, Graph reconstruction—A survey, *J. Graph Theory* **1** (1977) 227–268.

[Br41]   R. L. Brooks, On colouring the nodes of a network, *Math. Proc. Cambridge Philos. Soc.* **37** (1941) 194–197.

[Ca59]   P. Camion, Chemins et circuits hamiltoniens des graphes complets, *C. R. Acad. Sci. Paris Sér. A* **249** (1959) 2151–2152.

[C57]   A. Cayley, On the theory of the analytical forms called trees, *Philos. Mag.* (4) **13** (1857) 172–176 = *Math. Papers*, Vol. 3, 242–246.

[C81]   A. Cayley, On the analytical forms called trees, *Amer. J. Math.* **4** (1881) 266–268 = *Math. Papers*, Vol. 11, 365–376.

[C89]   A. Cayley, A theorem on trees, *Quart. J. Pure Appl. Math.* **23** (1889) 376–378 = *Math. Papers*, Vol. 13, 26–28.

[Ch71]   P. Z. Chinn, A graph with p points and enough distinct ($p - 2$)-order subgraphs is reconstructible, *Recent Trends in Graph Theory* (M. Capobianco et al., eds.) Lecture Notes in Math. 186, Springer, Berlin (1971) 71–73.

[ChG83]   F. R. K. Chung and C. M. Grinstead, A survey of bound for classical Ramsey numbers, *J. Graph Theory* **7** (1983) 25–37.

[Ch72]   V. Chvátal, On Hamilton's ideals, *J. Combin. Theory Ser. B* **12** (1972) 163–168.

[ChE72]   V. Chvátal and P. Erdös, A note on hamiltonian circuits, *Discrete Math.* **2** (1972) 111–113.

[Cl58]   L. E. Clarke, On Cayley's formula for counting trees, *J. London Math. Soc.* **33** (1958) 471–475.

[Co-U]   J. E. Cohen, Threshold phenomena in random structures, Unpublished.

[Co82]   J. E. Cohen, The asymptotic probability that a random graph is a unit interval graph, indifference graph, or proper interval graph, *Discrete Math.* **40** (1982) 21–24.

[CoKM79]   J. E. Cohen, J. Komlós and T. Mueller, The probability of an interval graph and why it matters, *Relations between combinatorics and other parts of mathematics* (D. K. Ray-Chaudhuri, ed.) Amer. Math. Soc., Providence (1979) 97–115.

[Di52]   G. A. Dirac, Some theorems on abstract graphs, *Proc. London Math. Soc. Ser. 3* **2** (1952) 69–81.

[Di60]   G. A. Dirac, Généralisation der théorème de Menger, *C.R. Acad. Sci. Paris* **250** (1960) 4252–4253.

[DiW83] J. D. Dixon and H. S. Wilf, The random selection of unlabelled graphs, *J. Algorithms* **4** (1983) 205–213.

[Er47] P. Erdös, Some remarks on the theory of graphs, *Bull. Amer. Math. Soc.* **53** (1947) 292–294.

[Er59] P. Erdös, Graph theory and probability, *Canad. J. Math.* **11** (1959) 34–38.

[ErF81] P. Erdös and S. Fajtlowicz, On the conjecture of Hajós, *Combinatorica* **1** (1981) 141–143.

[ErK-U] P. Erdös and J. W. Kennedy, *k*-Connectivity in random graphs, Unpublished.

[ErL75] P. Erdös and L. Lovász, Problems and results on 3-chromatic hypergraphs and some related questions, *Infinite and Finite Sets* (A. Hajnal et al., eds.) North-Holland (1975) 609–628.

[ErP83] P. Erdös and Z. Palka, Trees in random graphs, *Discrete Math.* **46** (1983) 145–150.

[ErP84] P. Erdös and Z. Palka, Addendum to Trees in random graphs, *Discrete Math.* **48** (1984) 331.

[ErPR83] P. Erdös, E. M. Palmer and R. W. Robinson, Local connectivity of a random graph, *J. Graph Theory* **7** (1983) 411–417.

[ErR59] P. Erdös and A. Rényi, On random graphs I, *Publ. Math. Debrecen* **6** (1959) 290–297.

[ErR60] P. Erdös and A. Rényi, On the evolution of random graphs, *Magyar Tud. Akad. Mat. Kutató Int. Közl.* **5** (1960) 17–61.

[ErR61] P. Erdös and A. Rényi, On the strength of connectedness of random graphs, *Acta Math. Acad. Sci. Hungar.* **12** (1961) 261–267.

[ErR66] P. Erdös and A. Rényi, On the existence of a factor of degree one of a connected random graph, *Acta Math. Acad. Sci. Hungar.* **17** (1966) 359–368.

[ErS74] P. Erdös and J. H. Spencer, *Probabilistic Methods in Combinatorics*, Academic, New York (1974).

[ErW77] P. Erdös and R. J. Wilson, On the chromatic index of almost all graphs, *J. Combin. Theory Ser. B* **23** (1977) 255–257.

[Et38] I. M. H. Etherington, On non-associative combinations, *Proc. Roy. Soc. Edinburgh* **59** (1938/39) 153–162.

[Fe57] W. Feller, *An Introduction to Probability Theory and Its Applications*, 2nd edn. Wiley, New York (1957).

[FeF82] T. I. Fenner and A. M. Frieze, On the connectivity of random *m*-orientable graphs and digraphs, *Combinatorica* **2** (1982) 347–359.

[FeF83] T. I. Fenner and A. M. Frieze, On the existence of hamiltonian cycles in a random graph, *Discrete Math.* **45** (1983) 301–305.

[FeF-U] T. I. Fenner and A. M. Frieze, Hamiltonian cycles in random regular graphs, *J. Combin. Theory Ser. B*.

[GaJ76] M. R. Garey and D. S. Johnson, The complexity of near-optimal graph coloring, *J. Assoc. Comput. Mach.* **23** (1976) 43–49.

[G56] E. N. Gilbert, Enumeration of labelled graphs, *Canad. J. Math.* **8** (1956) 405–411.

[G59] E. N. Gilbert, Random graphs, *Ann. Math. Stat.* **30** (1959) 1141–1144.

[Gl81] A. D. Glukhov, the maximal genus of a random graph, *Akad. Nauk Ukrain. SSR Inst. Mat. Preprint 1981, no. 2, Svoistva Grafov Nekotor. Klassov*, 20–22, 30–31.

[GrRS80] R. L. Graham, B. L. Rothschild, and J. H. Spencer, *Ramsey Theory*, Wiley, New York (1980).

[GrS71] R. L. Graham and J. H. Spencer, A constructive solution to a tournament problem, *Canad. Math. Bull.* **14** (1971) 45–48.

[Gr77]   G. R. Grimmett, Random graph theorems, *Trans. Seventh Prague Conf. on Information Theory and Related Topics* (A) (1977) 203–209.

[Gr80]   G. R. Grimmett, Random labelled trees and their branching networks, *J. Austral. Math. Soc.* (A) **30** (1980–81) 229–237.

[Gr83]   G. R. Grimmett, Random Graphs, *Selected Topics in Graph Theory* 2 (L. W. Beineke and R. J. Wilson, eds.) Academic, New York (1983) 201–225.

[GrM75] G. R. Grimmett and C. J. H. McDiarmid, On coloring random graphs, *Math. Proc. Cambridge Philos. Soc.* **77** (1975) 313–324.

[HaPR84] J. I. Hall, E. M. Palmer and R. W. Robinson, Redfield's lost paper in a modern context, *J. Graph Theory* **8** (1984) 225–240.

[H55a]   F. Harary, The number of linear, directed, rooted, and connected graphs, *Trans. Amer. Math. Soc.* **78** (1955) 445–463.

[H60]   F. Harary, Unsolved problems in the enumeration of graphs, *Magyar Tud. Akad. Mat. Kutató Int. Közl.* **5** (1960) 63–95.

[HP73]   F. Harary and E. M. Palmer, *Graphical Enumeration*, Academic, New York (1973).

[HP79]   F. Harary and E. M. Palmer, The probability that a point of a tree is fixed, *Math. Proc. Cambridge Philos. Soc.* **85** (1979) 407–415.

[HR84]   F. Harary and R. W. Robinson, The rediscovery of Redfield's papers, *J. Graph Theory* **8** (1984) 191–193.

[HRS75] F. Harary, R. W. Robinson and A. J. Schwenk, Twenty-step algorithm for determining the asymptotic number of trees of various species, *J. Austral. Math. Soc. Ser. A* **20** (1975) 483–503.

[Iv73]   G. I. Ivčhenko, On the asymptotic behavior of degrees of vertices in a random graph, *Theory Probab. Appl.* 18 (1973) 188–195.

[Ju81]   F. Juhasz, On the spectrum of a random graph, *Algebraic methods in graph theory*, Vol. I (L. Lovász and V. T. Sós, eds.) Colloq. Math. Soc. János Bolyai, North-Holland, Amsterdam (1981) 313–316.

[Ka82]   M. Karoński, A review of random graphs, *J. Graph Theory* **6** (1982) 349–389.

[Ka79]   R. M. Karp, Probabilistic analysis of a canonical numbering algorithm for graphs, *Relations between combinatorics and other parts of mathematics* (D. K. Ray-Chaudhuri, ed.) Amer. Math. Soc., Providence (1979) 365–378.

[Ka55]   L. Katz, Probability of indecomposability of a random mapping function, *Ann. Math. Statist.* **26** (1955) 512–517.

[Ke70]   A. K. Kel'mans, Bounds on the probability characteristics of random graphs, *Automat. Remote Control* **11** (1970) 1833–1839.

[Ke81a] A. K. Kel'mans, On graphs with randomly deleted edges, *Acta Math. Acad. Sci. Hungar.* **37** (1981) 77–88.

[Ke81]   J. W. Kennedy, Icycles—I, *The Theory and Applications of Graphs* (G. Chartrand et al., eds.) Wiley, New York (1981) 409–429.

[Ke83]   J. W. Kennedy, The random-graph like state of matter, *Computer Applications in Chemistry* (S. R. Heller and R. Potenzone, Jr., eds.) Elsevier, Amsterdam (1983) 151–178.

[KlL81] V. L. Klee and D. G. Larman, Diameters of random graphs, *Canad. J. Math.* **33** (1981) 618–640.

[KlLW80] V. L. Klee, D. G. Larman and E. M. Wright, The diameter of almost all bipartite graphs, *Studia Sci. Math. Hungar* **15** (1980) 39–43.

[KoSS80] J. Komlós, M. Sulyok and E. Szeméredi, Second largest component in a random graph, *Studia Sci. Math. Hungar.* **15** (1980) 391–395.

[KoS83]   J. Komlós and E. Szemerédi, Limit distribution for the existence of hamiltonian cycles in a random graph, *Discrete Math.* **43** (1983) 55–63.

[Ko76]   A. D. Koršunov, Solution of a problem of Erdös and Rényi on hamiltonian cycles in nonoriented graphs, *Soviet. Mat. Doklady* **17** (1976) 760–764.

[Ko72]   I. N. Kovalenko, Structure of an oriented random graph, *Teor. Verojatnost. i Mat. Statist.* **6** (1972) 83–91.

[Ku72]   W. W. Kuhn, A random graph generator, *Proc. 3rd S-E Conf. Combinatorics, Graph Theory, and Computing* (F. Hoffman et al, eds.) Utilitas, Winnipeg (1972) 311–313.

[LV76]   M. Las Vergnas, Sur les arborescences dans un graphe orienté *Discrete Math.* **15** (1976) 27–39.

[La76]   E. L. Lawler, *Combinatorial Optimization: Networks and Matroids*, Holt, New York (1976).

[Li-U]   R. J. Lipton, The beacon set approach to isomorphism, Unpublished report, Yale University (1978).

[Ll84]   E. K. Lloyd, J. Howard Redfield 1879–1944, *J. Graph Theory* **8** (1984) 195–203.

[Lo68]   L. Lovász, On chromatic number of finite set-systems, *Acta Math. Acad. Sci. Hungar.* **19** (1968) 59–67.

[Lo72]   L. Lovász, A note on the line reconstruction problem, *J. Combin. Theory Ser. B* **13** (1972) 309–310.

[MaS84]   M. M. Matthews and D. P. Sumner, Hamiltonian results in $K_{1,3}$-free graphs, *J. Graph Theory* **8** (1984) 139–176.

[Ma70]   D. W. Matula, On the complete subgraphs of a random graph, *Proc. 2nd Chapel Hill Conf. Combinatorial Math. and its Applications* (R. C. Bose et al., eds) Univ. North Carolina, Chapel Hill (1970) 356–369.

[Ma72]   D. W. Matula, The employee party problem, *Notices Amer. Math. Soc.* **19** (1972) A-382.

[Ma76]   D. W. Matula, The largest clique size in a random graph, Technical Report, Dept. of Computer Science, Southern Methodist University Dallas (1976).

[MaMI72]   D. W. Matula, G. Marble and J. D. Isaacson, Graph coloring algorithms, *Graph Theory and Computing* (R. C. Read, ed.) Academic, New York (1972) 109–122.

[Mc81]   C. J. H. McDiarmid, General percolation and random graphs, *Adv. Appl. Probab.* **13** (1981) 40–60.

[Mc82]   C. J. H. McDiarmid, Achromatic numbers of random graphs, *Math. Proc. Cambridge Philos. Soc.* **92** (1982) 21–28.

[Mc-U]   C. J. H. McDiarmid, Colouring random graphs, *Proc. Gargnano Conf. on Stochastics and Optimization*, Italy, 1982.

[McW-U]   B. D. McKay and N. C. Wormald, Automorphisms of random graphs with specified degrees, *Combinatorica*.

[MeM84]   A. Meir and J. W. Moon, On random mapping patterns, *Combinatorica* **4** (1984) 61–70.

[Me27]   K. Menger, Zur allgemeinen Kurventheorie, *Fund. Math.* **10** (1927) 96–115.

[M67]   J. W. Moon, Various proofs of Cayley's formula for counting trees, *A Seminar on Graph Theory* (F. Harary, ed.) Holt, New York (1967) 70–78.

[M70]   J. W. Moon, *Counting Labelled Trees*, Canada, Math. Congress, Montreal (1970).

[Mu76]   V. Müller, Probabilistic reconstruction from subgraphs, *Comment. Math. Univ. Carolin.* **17** (1976) 709–719.

[Mu77]   V. Müller, The edge reconstruction hypothesis is true for graphs with more than $n \log_2 n$ edges, *J. Combin. Theory Ser. B* **22** (1977) 281–283.

[NiW78]   A. Nijenhuis and H. S. Wilf, *Combinatorial algorithms for computers and calculators*, 2nd ed., Academic, New York (1968).

[ObS79]   D. Oberly and D. P. Sumner, Every connected, locally connected nontrivial graph with no induced claw is Hamiltonian, *J. Graph Theory* **3** (1979) 351–356.

[Ob67]   W. Oberschelp, Kombinatorische Anzahlbestimmungen in Relationen, *Math. Ann.* **174** (1967) 53–78.

[Or60]   O. Ore, Note on hamilton circuits, *Amer. Math. Monthly* **67** (1960) 55.

[O48]   R. Otter, the number of trees, *Ann. of Math.* **49** (1948) 583–599.

[Pal81]   Z. Palka, On pendant vertices in random graphs, *Colloq. Math.* **45** (1981) 159–167.

[Pal82a]   Z. Palka, Isolated trees in a random graph, *Zastos. Mat.* **17** (1982) 309–316.

[Pal82b]   Z. Palka, Isolated trees in a bichromatic random graph, *Colloq. Math.* **47** (1982) 153–162.

[Pal84]   Z. Palka, On the number of vertices of given degree in a random graph, *J. Graph Theory* **8** (1984) 167–170.

[Pal-U]   Z. Palka, Degree of vertices in a random graph, Unpublished.

[PaR-U]   E. M. Palmer and R. W. Robinson, Connectivity of a random *m*-graph, *Rev. Roum. Math. Pures et Appl.*

[PaR-P]   E. M. Palmer and R. W. Robinson, Asymptotic number of symmetries in locally restricted trees, In preparation.

[Pa81]   V. Palmer, On the connectivity of regular random digraphs, *Utilitas Math.* **20** (1981) 293–299.

[Pi82]   B. Pittel, On the probable behavior of some algorithms for finding the stability number of a graph, *Math. Proc. Cambridge Philos. Soc.* **92** (1982) 511–526.

[P37]   G. Pólya, Kombinatorische Anzahlbestimmungen für Gruppen, graphen und chemische Verbindungen, *Acta Math.* **68** (1937) 145–254.

[Po62]   L. Pósa, A theorem concerning hamilton lines, *Magyar Tud. Akad. Mat. Kutató Int. Közl.* **7** (1962) 225–226.

[Po76]   L. Pósa, Hamiltonian circuits in random graphs, *Discrete Math.* **14** (1976) 359–364.

[Pr18]   H. Prüfer, Neuer Beweis eines Satzes uber Permutationen, *Arch. Math. Phys.* **27** (1918) 742–744.

[Re59]   R. C. Read, The enumeration of locally restricted graphs, I and II, *J. London Math. Soc.* **34** (1959) 417–436; **35** (1960) 334–351.

[Re70]   R. C. Read, Some unusual enumeration problems, *Ann. N.Y. Acad. Sci.* **175** (1970) 314–326.

[Re78]   R. C. Read, Some applications of computers in graph theory, *Selected Topics in Graph Theory*, Academic, New York (1978) 417–444.

[ReC77]   R. C. Read and D. G. Corneil, The graph isomorphism disease, *J. Graph Theory* **1** (1977) 339–363.

[ReW80]   R. C. Read and N. C. Wormald, Number of labeled 4-regular graphs, *J. Graph Theory* **4** (1980) 203–212.

[R27]   J. H. Redfield, The theory of group-reduced distributions, *Amer. J. Math.* **49** (1927) 433–455.

[Re59a]   A. Rényi, Some remarks on the theory of trees, *Magyar Tud. Akad. Mat. Kutató Int. Közl.* **4** (1959) 73–85.

[Re59b]   A. Rényi, On connected graphs. I, *Magyar Tud. Akad. Mat. Kutató Int. Közl.* **4** (1959) 385–388.

[Re70]   A. Rényi, *Probability Theory*, North-Holland, Amsterdam (1970).

[Ri51] R. J. Riddell, *Contributions to the theory of condensation*, Dissertation, Univ. of Michigan, Ann Arbor (1951).

[RiU53] R. J. Riddell and G. E. Uhlenbeck, On the theory of the virial development of the equation of state of monoatomic gases, *J. Chem. Phys.* **21** (1953) 2056–2064.

[Ri58] J. Riordan, *An Introduction to Combinatorial Analysis*, Wiley, New York (1958).

[Ro77b] R. W. Robinson, Counting cubic graphs, *J. Graph Theory* **1** (1977) 285–286.

[RoS75] R. W. Robinson and A. J. Schwenk, The distribution of degrees in a large random tree, *Discrete Math.* **12** (1975) 359–372.

[RoW84] R. W. Robinson and N. C. Wormald, Existence of long cycles in random cubic graphs, *Enumeration and Design*, Academic, Canada (1984) 251–270.

[Ro81] S. M. Ross, A random graph, *J. Appl. Probab.* **18** (1981) 309–315.

[Ru81] A. Ruciński, The *r*-connectedness of *k*-partite random graph, *Bull. Acad. Polon. Sci. Ser. Sci. Math.* **29** (1981) 321–330.

[Sc79] K. Schürger, Limit theorems for complete subgraphs of random graphs, *Period. Math. Hungar.* **10** (1979) 47–53.

[Sc73] A. J. Schwenk, Almost all trees are cospectral, *New Directions in the Theory of Graphs* (F. Harary, ed.) Academic, New York (1973) 275–307.

[Sh83] E. Shamir, How many random edges make a graph hamiltonian? *Combinatorica* **3** (1983) 123–131.

[ShU81] E. Shamir and E. Upfal, On factors in random graphs, *Israel J. Math.* **39** (1981) 296–302.

[ShU82] E. Shamir and E. Upfal, One-factor in random graphs based on vertex choice, *Discrete Math.* **41** (1982) 281–286.

[Sp78] J. H. Spencer, Nonconstructive methods in discrete mathematics, *Studies in Combinatorics* (G.-C. Rota, ed.) Math. Assoc. Amer., (1978) 142–178.

[St77] P. K. Stockmeyer, The falsity of the reconstruction conjecture for tournaments, *J. Graph Theory* **1** (1977) 19–25.

[St81] P. K. Stockmeyer, Which reconstruction results are significant? *The Theory and Applications of Graphs* (G. Chartrand et al., ed.) Wiley, New York (1981) 543–555.

[Th78] K.-J. Thürlings, Direkte Berechnung spezieller Anzahlen in Pólyas Abzähltheorie, *Match* No. 4 (1978) 161–171.

[Ti79] G. Tinhofer, On the generation of random graphs with given properties and known distributions, *Appl. Comput. Sci., Ber. Prakt. Inf.* **13** (1979) 265–297.

[Ti80] G. Tinhoffer, Zufallsgraphen (Random Graphs), *Appl. Comput. Sci., Ber. Prakt. Inf.* **17**, Carl Hanson, Munich (1980).

[To80] E. Toman, Probability of connectedness of a random subgraph of an *n*-dimensional cube, *Math. Slovaca* **30** (1980) 251–265.

[To81] I. Tomescu, On the chromatic number of almost graphs, *Bull. Math. Soc. Sci. Math. R.S. Roumanie* (*N.S.*) **25** (73) (1981) 321–323.

[T47] W. T. Tutte, The factorizations of linear graphs, *J. London Math. Soc.* **22** (1947) 107–111.

[Vi64] V. G. Vizing, On an estimate of the chromatic class of a *p*-graph (Russian), *Diskret. Analiz* **3** (1964) 25–30.

[Vi65] V. G. Vizing, Critical graphs with a given chromatic class (Russian), *Diskret. Analiz* **5** (1965) 9–17.

[Wa79] D. W. Walkup, On the expected value of a random assignment problem, *SIAM J. Comput.* **8** (1979) 440–442.

[Wa80]   D. W. Walkup, Matchings in random regular bipartite digraphs, *Discrete Math.* **31** (1980) 59–64.

[Wh32]   H. Whitney, Congruent graphs and the connectivity of graphs, *Amer. J. Math.* **54** (1932) 150–168.

[Wi81]   H. S. Wilf, The uniform selection of free trees, *J. Algorithms* **2** (1981) 204–207.

[Wo78c] N. C. Wormald, *Some problems in the enumeration of labelled graphs*, Dissertation, Univ. of Newcastle, N.S.W. (1978).

[Wo79b] N. C. Wormald, Enumeration of labelled graphs II: Cubic graphs with a given connectivity, *J. London Math. Soc.* (2) **20** (1979) 1–7.

[Wo81c] N. C. Wormald, The asymptotic connectivity of labelled regular graphs, *J. Combin. Theory Ser. B* **31** (1981) 156–167.

[Wo81d] N. C. Wormald, The asymptotic distribution of short cycles in random graphs, *J. Combin. Theory Ser. B* **31** (1981) 168–182.

[Wo84]   N. C. Wormald, Generating random regular graphs, *J. Algorithms* **5** (1984) 247–280.

[W70a]   E. M. Wright, Asymptotic enumeration of connected graphs, *Proc. Roy. Soc. Edinburgh Sect. A* **68** (1968/70) 298–308.

[W70b]   E. M. Wright, Graphs on unlabelled nodes with a given number of edges, *Acta Math.* **126** (1970) 1–9.

[W72a]   E. M. Wright, The probability of connectedness of an unlabelled graph can be less for more edges, *Proc. Amer. Math. Soc.* **35** (1972) 21–25.

[W74a]   E. M. Wright, Graphs on unlabelled nodes with a large number of edges, *Proc. London Math. Soc.* (3) **28** (1974) 577–594.

[W74b]   E. M. Wright, For how many edges is a graph almost certainly hamiltonian?, *J. London Math. Soc.* (2) **8** (1974) 44–48.

[W74c]   E. M. Wright, Asymmetric and symmetric graphs, *Glasgow Math. J.* **15** (1974) 69–73.

[W74d]   E. M. Wright, Two problems in the enumeration of unlabelled graphs, *Discrete Math.* **9** (1974) 289–292.

[W75a]   E. M. Wright, The probability of connectedness of a large unlabelled graph, *J. London Math. Soc.* (2) **11** (1975) 13–16.

[W76a]   E. M. Wright, The evolution of unlabelled graphs, *J. London Math. Soc.* (2) **14** (1976) 554–558.

[W76b]   E. M. Wright, The asymptotic enumeration of unlabelled graphs, *Proc. 5th British Comb. Conf. 1975* (C. St. J.A. Nash-Williams and J. Sheehan, eds.) Utilitas, Winipeg (1976) 665–677.

[W76c]   E. M. Wright, The proportion of unlabelled graphs which are hamiltonian, *Bull. London Math. Soc.* **8** (1976) 241–244.

[W82]   E. M. Wright, The *k*-connectedness of bipartite graphs, *J. London Math. Soc.* (2) **25** (1982) 7–12.

[W83]   E. M. Wright, The number of sparsely edged labelled hamiltonian graphs, *Glasgow Math. J.* **24** (1983) 83–87.

NAME INDEX

I would never join a club that
would admit me as a member.

GROUCHO MARX

173

SUBJECT INDEX

Anything you look for in the Yellow
Pages will not be listed in the
category you first try to find it
under. Start with the second.

ANDY ROONEY

*Ironically, fish do not obey the law but, as observed by Feller [Fe57] and confirmed by research at MSU, cowflops do.

OTHER TITLES BY
THE AUTHOR

Look Homeward, Angle

Great Expectation

The Texas Chainrule Massacre

Everything You Always Wanted To Know About X*

Joy of Arithmetic

*But were afraid to ask

It ain't over 'til it's over

YOGI BERRA

*(From the inscription on the tomb
of the unknown professor at
Michigan State University.)*